FOURIER SERIES
AND
PARTIAL DIFFERENTIAL EQUATIONS

A SERIES OF PROGRAMMES ON DIFFERENTIAL EQUATIONS

CONSULTANT EDITOR:

A. C. BAJPAI

PROFESSOR OF MATHEMATICAL EDUCATION AND DIRECTOR OF CAMET

GENERAL EDITORS

I. M. CALUS

J. A. FAIRLEY

CAMET

(CENTRE FOR THE ADVANCEMENT OF MATHEMATICAL EDUCATION IN TECHNOLOGY)

DEPARTMENT OF MATHEMATICS

LOUGHBOROUGH UNIVERSITY OF TECHNOLOGY

FOURIER SERIES
AND
PARTIAL DIFFERENTIAL EQUATIONS

A PROGRAMMED COURSE FOR STUDENTS OF SCIENCE AND TECHNOLOGY

I. M. CALUS
J. A. FAIRLEY

LOUGHBOROUGH UNIVERSITY OF TECHNOLOGY

WILEY - INTERSCIENCE
A DIVISION OF
JOHN WILEY & SONS LTD
LONDON · NEW YORK · SYDNEY · TORONTO

Copyright © 1970 by John Wiley & Sons Ltd., All Rights Reserved. No part of this book may be reproduced stored in a retrieval system, or transmitted in any form or by any means, electronic, mechanical photo copying, recording or otherwise without the prior written permission of the copyright owner

Library of Congress Catalog Card No. 72-140893

ISBN 0 471 13070 2

Printed photo litho in Great Britain by Page Bros. (Norwich) Ltd.

EDITORS' PREFACE

These two programmes are designed to meet the needs of scientists and technologists who find that their work requires a knowledge of Fourier series and partial differential equations. Consequently the emphasis is on the practical side of the subject in each case, and the more theoretical aspects have been omitted.

The programmed method of presentation, used throughout, has many advantages. The development of the subject progresses in carefully sequenced steps, with the student proceeding at his own pace. At each stage he has an active part to play by answering a question or solving a problem, and thus learns by doing. By comparing his own answer with that given in the text, he obtains a continuous assessment of his understanding of the subject up to that point. Explanation of the material covered is given in greater detail than is usually to be found in conventional style textbooks, especially at points where, in their experience as teachers, the authors find that students often have difficulty.

In spite of careful checking by the general editors, it is possible that the occasional error has crept through. They would appreciate receiving information about any such mistakes that might be discovered.

A debt of gratitude to the following is acknowledged with pleasure:

Loughborough University of Technology for supporting this venture.

Staff and students of Loughborough University of Technology and other institutions who have participated in the testing of these programmes.

Our Colleague, Mr. E. Besag, of the Department of Electronic and Electrical Engineering, for his advice on applications in his field.

Mrs. June Russell for preparing the camera-ready copy from which the book has been printed.

John Wiley and Sons Ltd for their help and cooperation.

CONTENTS

1. FOURIER SERIES — 1:1
2. PARTIAL DIFFERENTIAL EQUATIONS FOR TECHNOLOGISTS — 2:1

FOURIER SERIES

A PROGRAMMED TEXT

I. M. Calus
J. A. Fairley

INSTRUCTIONS

This programme constitutes a self-instructional course on Fourier Series. The programme is divided up into a number of FRAMES which are to be worked *in the order given*. You will be required to participate in many of these frames and in such cases the answers are provided in ANSWER FRAMES, designated by the letter A following the frame number. Steps in the working are given where this is considered helpful. The answer frame is separated from the main frame by a line of asterisks: ****************** Keep the answers covered until you have written your own response.

If your answer is wrong, go back and see why. Do not proceed to the next frame until you have corrected any mistakes in your attempt and are satisfied that you understand the contents up to this point.

Before reading this programme it is necessary that you are familiar with the following

Prerequisites

Standard formulae in Trigonometry such as
$\sin A \cos B = \frac{1}{2}[\sin(A+B) + \sin(A-B)]$ and the expression of $a \cos \theta + b \sin \theta$ in the form $r \sin(\theta + \alpha)$.

Integration of powers of x, trigonometric functions involving sines and cosines, and integration by parts.

Mean values.

FOURIER SERIES

FRAME 1

Introduction

You have probably already met examples of the use of Maclaurin's series where a function f(x) is expressed as a series of powers of x,

$$\text{e.g.} \quad \log_e(1 + x) = x - \frac{x^2}{2} + \frac{x^3}{3} - \frac{x^4}{4} + \ldots$$

We do this because it is often easier to deal with powers of x than with the original functions.

In some problems, e.g. those dealing with oscillations, it is more convenient to use a series of sines. As sines are periodic, such a series can only represent a periodic function. In the next frame we shall remind you of some of the features of periodic functions.

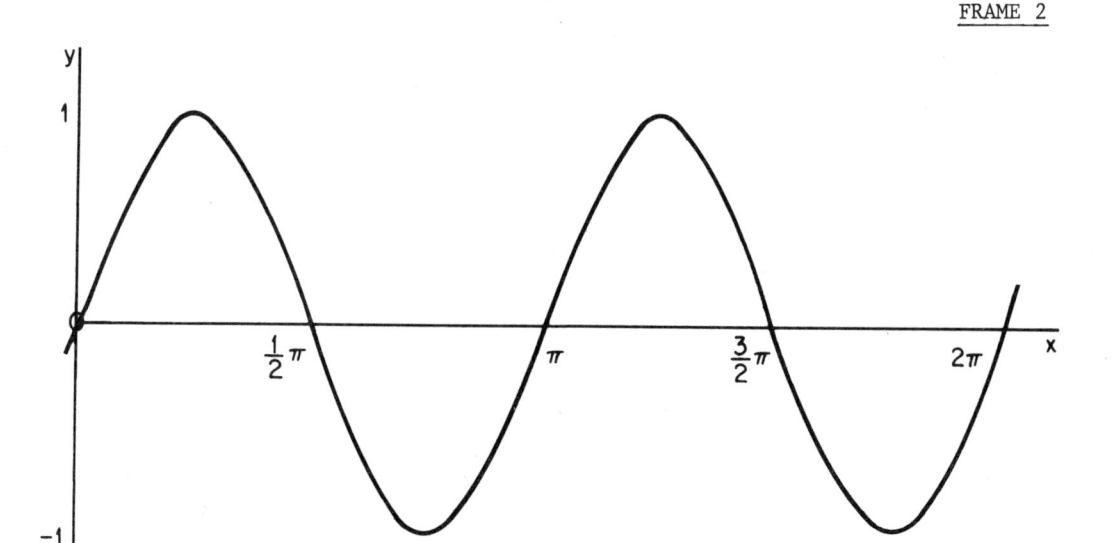

FRAME 2

This is the graph of y = sin 2x. The period is π.

Note that sin 2x = sin 2(x + π) for all values of x, i.e. increasing x by π does not change the value of y.

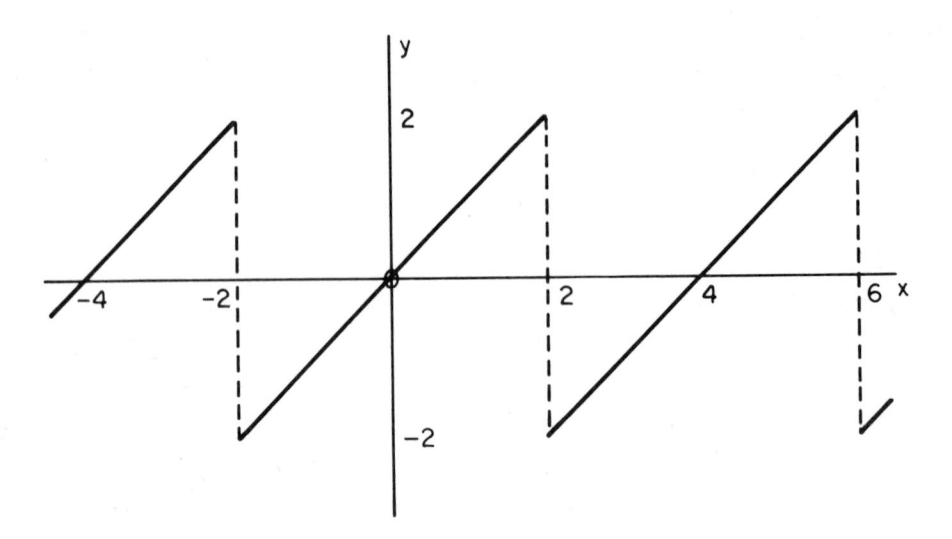

Here the period is 4 and the graph is of y = f(x) where

$$f(x) = x \qquad -2 < x < +2$$

and f(x + 4) = f(x)   for all x.

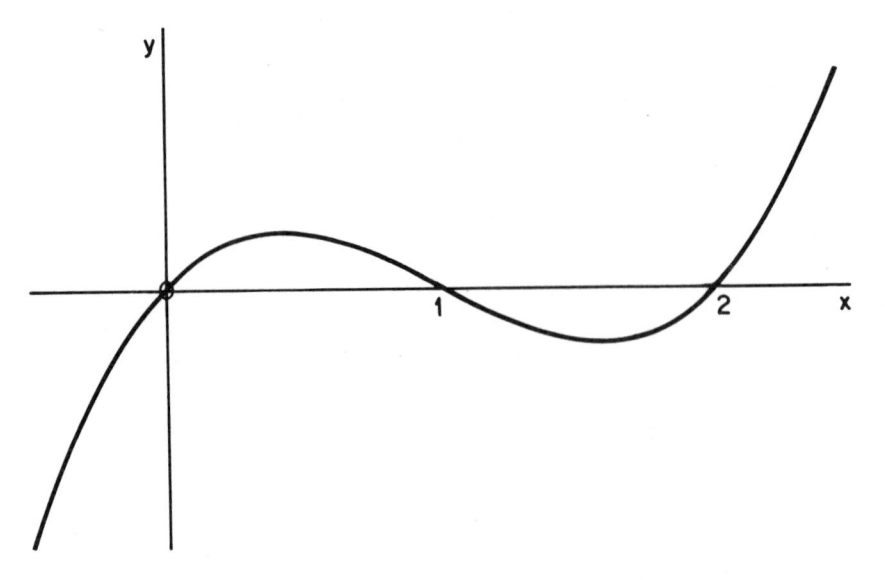

This is the graph of y = x(x − 1)(x − 2) and in this case the function is not periodic.

FOURIER SERIES

FRAME 2 continued

State whether each of the following is periodic. If it is, give the period and a set of equations which define the function.

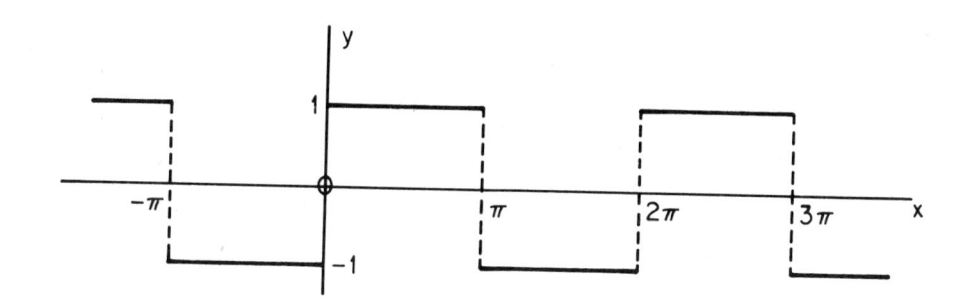

(i)

(ii)

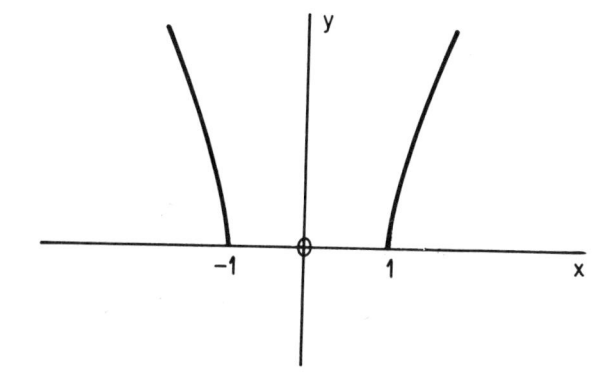

(iii)

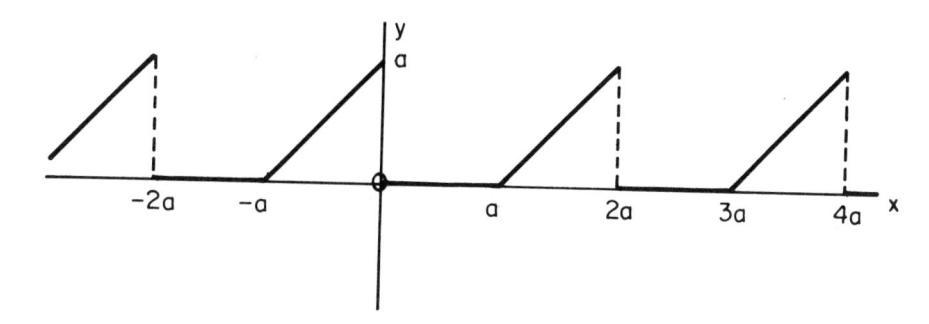

****

2A

(i) *Periodic.*

Period 2π.

The most obvious are

$$\begin{cases} f(x) = 1 & 0 < x < \pi \\ f(x) = -1 & \pi < x < 2\pi \\ f(x + 2\pi) = f(x) \end{cases} \quad \underline{or} \quad \begin{cases} f(x) = -1 & -\pi < x < 0 \\ f(x) = 1 & 0 < x < \pi \\ f(x + 2\pi) = f(x) \end{cases}$$

but there are, of course, many other possibilities.

(ii) *Not periodic.*

(iii) *Periodic.*

Period $2a$.

$$\begin{cases} f(x) = 0 & 0 < x < a \\ f(x) = x - a & a < x < 2a \\ f(x + 2a) = f(x) \end{cases} \quad \underline{or} \quad \begin{cases} f(x) = x + a & -a < x < 0 \\ f(x) = 0 & 0 < x < a \\ f(x + 2a) = f(x) \end{cases}$$

Again, there are other possible sets of equations.

FRAME 3

A periodic function $f(x)$, whose period is 2π, can be expressed in the form

$$f(x) = c_0 + c_1 \sin(x + \alpha_1) + c_2 \sin(2x + \alpha_2) + c_3 \sin(3x + \alpha_3) + \ldots \quad (3.1)$$

provided $f(x)$ satisfies certain other conditions. For functions which you are likely to meet, the only conditions which you will need to ensure are satisfied are:

(i) the function must be single-valued, i.e. for each value of x there must be only one value of $f(x)$,

(ii) the function must never be infinite.

The term $c_1 \sin(x + \alpha_1)$ is called the first harmonic or fundamental.

The terms $c_2 \sin(2x + \alpha_2)$, $c_3 \sin(3x + \alpha_3)$ etc. are called the second harmonic, third harmonic, etc. A series such as that on the R.H.S. of (3.1) is called a FOURIER SERIES.

FOURIER SERIES

FRAME 3 continued

If the period of $f(x)$ is not 2π, a conversion to a function whose period is 2π can be made by a suitable change of variable, as will be shown later in the programme.

FRAME 4

The series in (3.1) can be expressed in an alternative form as follows:

$$\sin(x + \alpha_1) = \sin x \cos \alpha_1 + \cos x \sin \alpha_1$$

and so we can write $\quad c_1 \sin(x + \alpha_1) = a_1 \cos x + b_1 \sin x$

where $\quad a_1 = c_1 \sin \alpha_1 \quad$ and $\quad b_1 = c_1 \cos \alpha_1$.

Converting the other terms in a similar way, the series becomes

$$f(x) = \tfrac{1}{2}a_0 + a_1 \cos x + a_2 \cos 2x + a_3 \cos 3x + \ldots$$
$$+ b_1 \sin x + b_2 \sin 2x + b_3 \sin 3x + \ldots \quad (4.1)$$

The reason for writing c_0 as $\tfrac{1}{2}a_0$ will become clear later on.

The problem now, in any particular case, is to find the coefficients $a_0, a_1, a_2, \ldots, b_1, b_2, \ldots$, or, what is equivalent, the values of $c_0, c_1, c_2, \ldots, \alpha_1, \alpha_2, \ldots$. In practice, it is easier to calculate the a's and the b's, and the c's and the α's can then be found from them, if the series is required in the form (3.1). The form (4.1) may, however, be suitable in itself.

Before proceeding to the method of calculating the coefficients, we shall, in the next few frames, give some illustrations showing that the idea of such a series for a periodic function is feasible and how it can be useful to an engineer or applied scientist.

FRAME 5

Let us consider the series

$$\frac{1}{2} + \frac{2}{\pi}(\sin x + \frac{1}{3}\sin 3x + \frac{1}{5}\sin 5x + \frac{1}{7}\sin 7x + \ldots) \qquad (5.1)$$

and draw the graphs representing:

(i) $\quad y = \frac{1}{2}$

(ii) $\quad y = \frac{1}{2} + \frac{2}{\pi}\sin x$

(iii) $\quad y = \frac{1}{2} + \frac{2}{\pi}(\sin x + \frac{1}{3}\sin 3x)$

(iv) $\quad y = \frac{1}{2} + \frac{2}{\pi}(\sin x + \frac{1}{3}\sin 3x + \frac{1}{5}\sin 5x)$

These graphs, from $x = 0$ to 2π, are:

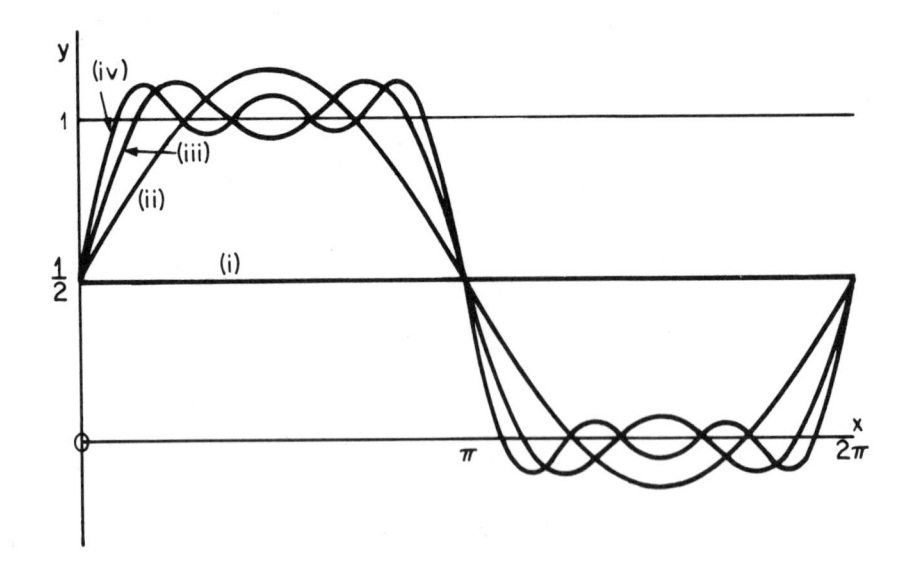

As (ii), (iii) and (iv) are periodic, each with period 2π, the continuation of their graphs for other values of x will simply be repetitions of the parts we have shown.

You will notice that, as the number of terms is increased, the oscillations about $y = 1$ and $y = 0$ get smaller and take place over a greater width. If we now considerably increase the number of terms in the sum, the graph will look like this:

FOURIER SERIES

FRAME 5 continued

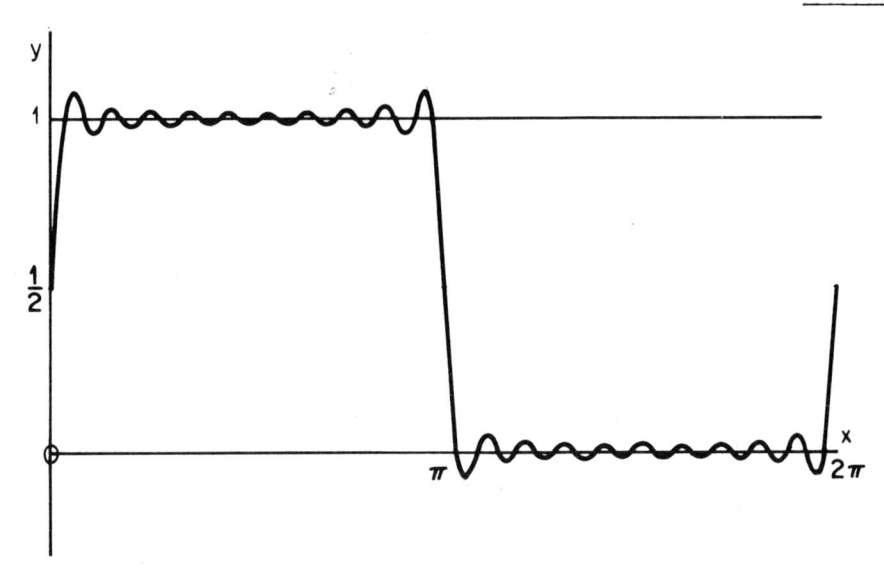

It is therefore reasonable to suggest that if we increase the number of terms indefinitely, the square-wave function shown will be produced.

This has equations
$$\begin{cases} f(x) = 1 & 0 < x < \pi \\ f(x) = 0 & \pi < x < 2\pi \\ f(x + 2\pi) = f(x) \end{cases}$$

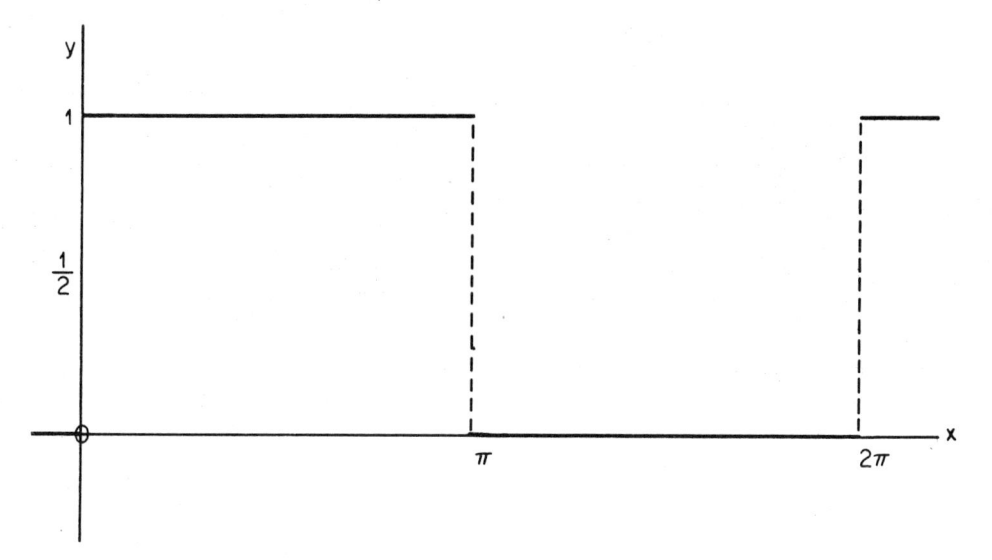

FOURIER SERIES

FRAME 5 continued

Later in the programme we shall show that the Fourier series for this square-wave function is the series (5.1).

You will notice that, as the series progresses, the coefficients decrease in magnitude. This fact is often made use of in engineering problems where approximations are made by neglecting the higher harmonics.

(The series (5.1) is rather special in that $a_1 = a_2 = a_3 = \ldots = 0$, but this does not affect the argument. We chose such a case merely to simplify the sketching of the graphs.)

FRAME 6

Some methods used by engineers in solving problems are only applicable to sinusoidal functions and consequently they often find it helpful to use Fourier series to express other functions in terms of sines (and/or cosines). In the field of electrical engineering two such cases are:

(i) The use of complex numbers in dealing with a.c. circuits depends on functions being sinusoidal. For example, if a voltage $v = V \cos \omega t = \text{Re}(Ve^{j\omega t})$, where Re denotes "real part of", is applied to a pure inductance L, the current is given by

$$\begin{aligned} i &= \text{Re}\{\tfrac{1}{j\omega L} Ve^{j\omega t}\} \\ &= \text{Re}\{\tfrac{V}{\omega L} e^{j(\omega t - \frac{1}{2}\pi)}\} \\ &= \tfrac{V}{\omega L} \cos(\omega t - \tfrac{\pi}{2}) \quad\quad (6.1) \end{aligned}$$

If a voltage is such that it can be written in the form

$$v = V_1 \cos \omega t + V_2 \cos 2\omega t + V_3 \cos 3\omega t + \ldots$$

i.e. as a Fourier series, then the corresponding current is

$$i = \tfrac{V_1}{\omega L} \cos(\omega t - \tfrac{\pi}{2}) + \tfrac{V_2}{2\omega L} \cos(2\omega t - \tfrac{\pi}{2}) + \tfrac{V_3}{3\omega L} \cos(3\omega t - \tfrac{\pi}{2}) + \ldots$$

by applying the result (6.1) to each term.

FOURIER SERIES

FRAME 6 continued

(ii) If the average value of the power in a circuit over time T is required, T being the period of the applied voltage, we have to evaluate

$$\frac{1}{T} \int_0^T vi \, dt$$

The integration involved is simplified if we can write the voltage and current in the forms

$$v = V_1 \cos(\omega t + \alpha_1) + V_2 \cos(2\omega t + \alpha_2) + \ldots$$

$$i = I_1 \cos(\omega t + \alpha_1 + \phi_1) + I_2 \cos(2\omega t + \alpha_2 + \phi_2) + \ldots$$

as then the integrals of all terms of the form

$$V_m \cos(m\omega t + \alpha_m) I_n \cos(n\omega t + \alpha_n + \phi_n), \text{ where } m \neq n, \text{ are zero.}$$

We shall be proving a simpler version of this result in FRAME 13. Consequently, we are only left with integrals of terms of the form

$$V_n \cos(n\omega t + \alpha_n) I_n \cos(n\omega t + \alpha_n + \phi_n)$$

which are very simple to evaluate.

You will notice that in these two cases, t is used instead of x as the independent variable is time. This is usually the case in electrical examples, but in other applications the independent variable is distance, generally represented by x. For uniformity within the programme, we shall work in terms of x.

You will also notice that multiples of ωt, rather than t, appear in these series. This is because the period of the waveforms is $2\pi/\omega$ instead of 2π.

FRAME 7

Fourier series are also useful in tackling some beam problems, as in the examples shown here and in the next frames.

If the load on a beam varies sinusoidally so also does the deflection and a formula for this is known. If the load can be expressed as the sum of such

FRAME 7 continued

terms then this formula can be applied to each term in turn. One such loading for which this can be done is the so-called patch loading, i.e. loading of the form

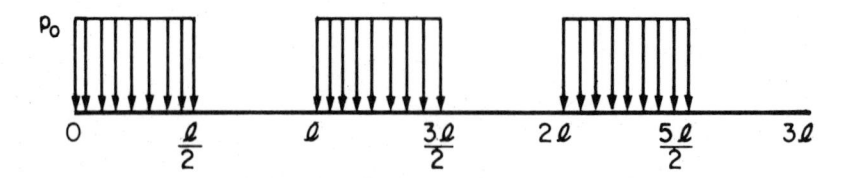

taking the length of the beam in this case as 3ℓ.

This can be expressed as a series of sine terms and later we shall see that it is

$$\frac{1}{2} P_o + \frac{2}{\pi} P_o \left(\sin \frac{2\pi x}{\ell} + \frac{1}{3} \sin \frac{6\pi x}{\ell} + \frac{1}{5} \sin \frac{10\pi x}{\ell} + \ldots \right) \qquad (7.1)$$

Multiples of $\frac{2\pi x}{\ell}$ occur here because the period is ℓ.

The calculation of the deflection due to the constant term is as straightforward as it is for the sine terms.

FRAME 8

Any value of x can be substituted in (7.1) but the practical interpretation only exists for $0 \leq x \leq 3\ell$ i.e. (7.1) represents the waveform

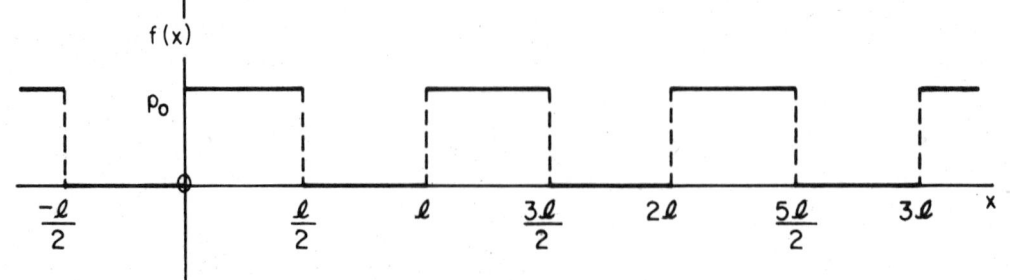

but we are only interested in the section between 0 and 3ℓ.

If the loading on a beam is not periodic it may still be helpful to be able to represent the deflection curve by means of a Fourier series.

FOURIER SERIES

FRAME 8 continued

Suppose, for example, we have a cantilever whose deflection curve is as in the diagram.

Some of the waveforms of which this is part are shown below.

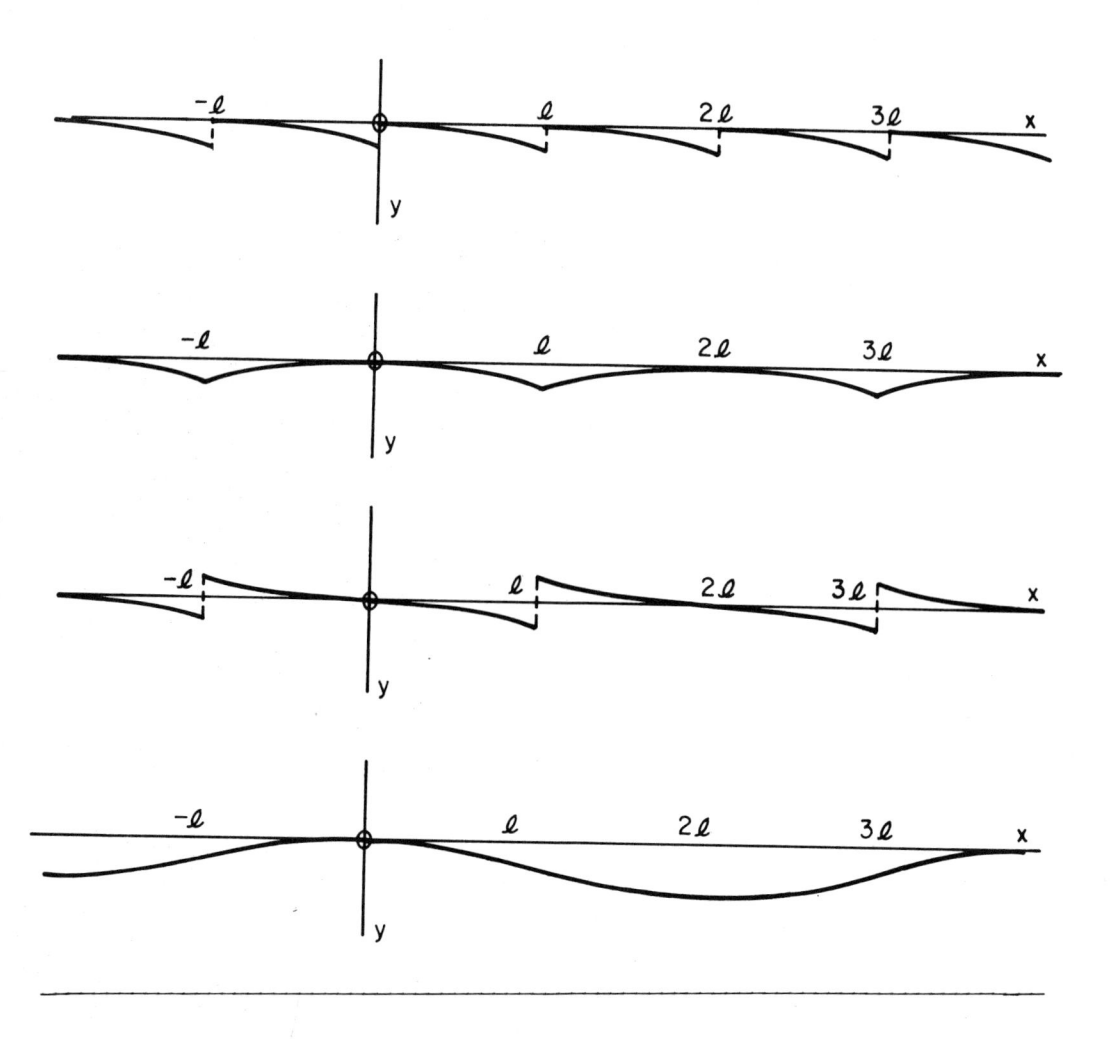

FRAME 9

You are probably familiar with the standard beam equations

$$EI \frac{d^2y}{dx^2} = M \quad \text{and} \quad EI \frac{d^4y}{dx^4} = w.$$

If the loading on the beam is irregular the solution of these equations may well not be a straightforward matter. An alternative approach is by a method using Fourier series. As a simple example, consider a cantilever loaded as shown.

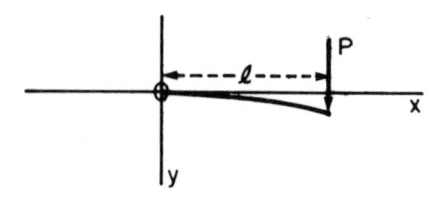

A solution for y in the form of a Fourier series can be found as follows: In the main part of this programme we shall show how to calculate coefficients of series for known functions. In the present problem y is not known as a function of x and the coefficients are found by considering certain physical aspects of the situation.

The conditions $y = \frac{dy}{dx} = 0$ at $x = 0$ and $\frac{d^2y}{dx^2} = 0$ at $x = \ell$ have to be satisfied, and this requirement is met by

$$y = A_1(1 - \cos \frac{\pi x}{2\ell}) + A_3(1 - \cos \frac{3\pi x}{2\ell}) + A_5(1 - \cos \frac{5\pi x}{2\ell}) + \ldots$$

(the last waveform in the previous frame).

You may like to check that this series does, in fact, satisfy the boundary conditions and you will then understand that the constant term appears as $A_1 + A_3 + A_5 + \ldots$ for convenience.

****

FOURIER SERIES

9A

$$\frac{dy}{dx} = A_1 \frac{\pi}{2\ell} \sin \frac{\pi x}{2\ell} + A_3 \frac{3\pi}{2\ell} \sin \frac{3\pi x}{2\ell} + A_5 \frac{5\pi}{2\ell} \sin \frac{5\pi x}{2\ell} + \ldots$$

$$\frac{d^2y}{dx^2} = A_1 \frac{\pi^2}{4\ell^2} \cos \frac{\pi x}{2\ell} + A_3 \frac{9\pi^2}{4\ell^2} \cos \frac{3\pi x}{2\ell} + A_5 \frac{25\pi^2}{4\ell^2} \cos \frac{5\pi x}{2\ell} + \ldots$$

which can be written as $y'' = C_1 \cos \frac{\pi x}{2\ell} + C_3 \cos \frac{3\pi x}{2\ell} + C_5 \cos \frac{5\pi x}{2\ell} + \ldots$

It is then easily verified that the conditions are satisfied.

FRAME 10

In this particular problem the coefficients are found by making use of the principle of virtual work. In order to do so, we require to know the energy stored in the beam. This is given by

$$U = \frac{EI}{2} \int_0^\ell (y'')^2 \, dx$$

$$= \frac{EI}{2} \int_0^\ell \left(C_1 \cos \frac{\pi x}{2\ell} + C_3 \cos \frac{3\pi x}{2\ell} + C_5 \cos \frac{5\pi x}{2\ell} + \ldots \right)^2 dx$$

from answer frame 9A.

As in the case of the average power example in FRAME 6, the integration is very simple because all the integrals

$$\int_0^\ell \cos \frac{\pi x}{2\ell} \cos \frac{3\pi x}{2\ell} \, dx, \quad \int_0^\ell \cos \frac{\pi x}{2\ell} \cos \frac{5\pi x}{2\ell} \, dx, \quad \int_0^\ell \cos \frac{3\pi x}{2\ell} \cos \frac{5\pi x}{2\ell} \, dx \quad \text{etc}$$

are zero, leaving only

$$\frac{EI}{2} \int_0^\ell \left(C_1^2 \cos^2 \frac{\pi x}{2\ell} + C_3^2 \cos^2 \frac{3\pi x}{2\ell} + C_5^2 \cos^2 \frac{5\pi x}{2\ell} + \ldots \right) dx$$

This is easily evaluated and the coefficients are then found by applying the principle of virtual work. The equation of the deflection curve thus obtained is

$$y = \frac{32P\ell^3}{EI\pi^4} \left\{ \left(1 - \cos \frac{\pi x}{2\ell}\right) + \frac{1}{3^4}\left(1 - \cos \frac{3\pi x}{2\ell}\right) + \frac{1}{5^4}\left(1 - \cos \frac{5\pi x}{2\ell}\right) + \ldots \right\}$$

FRAME 11

Another application of Fourier series is their use in fitting boundary conditions to the solutions of partial differential equations.

One example is the case of a taut string, one point of which is pulled aside and then released. The string oscillates and the displacement of any point depends both on its distance from one end and on time. The equation of motion is

$$\frac{\partial^2 y}{\partial x^2} = \frac{1}{c^2} \frac{\partial^2 y}{\partial t^2}$$

where c is a constant and y is the displacement of a point distant x from one end at time t. This is a partial differential equation as y is a function of both x and t. One of the boundary conditions is that when $t = 0$ the shape of the string is

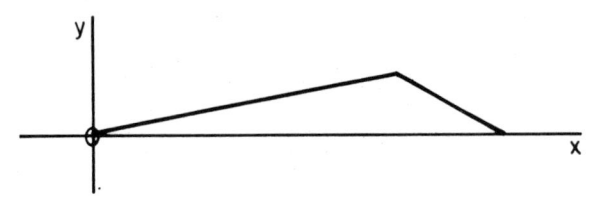

and it is found that in order to apply this boundary condition it is necessary to express this shape as a Fourier series.

Other examples of practical problems requiring a similar technique are certain cases of rod oscillations, heat conduction and voltages in an electric cable. The solution of such partial differential equations is dealt with in the next programme in this volume.

FRAME 12

Finally we would mention one or two applications of Fourier series in medical science.

A musical note consists of a fundamental and higher harmonics, the latter being the overtones, and, in the study of the physiology of hearing, Helmholtz's theory assumes that different fibres in the inner ear are excited only by those

FOURIER SERIES

FRAME 12 continued

harmonics to which they are tuned.

An electric current is produced by the activity of the heart. The variation of the voltage involved is periodic and can be measured by an electrocardiograph. The resulting waveform can be analysed into a Fourier series and the coefficients will show marked changes if disease or damage is present.

Human temperature fluctuates periodically, and can thus be represented by a Fourier series (see FRAME 53 Miscellaneous Examples, No. 8).

FRAME 13

Integrals required for calculating the coefficients

The calculation of the coefficients a_0, a_1, a_2, ..., b_1, b_2, ... will require the values of certain definite integrals, which we will now find. In these integrals both m and n are integers.

Two of the integrals that will be required are $\int_{-\pi}^{\pi} \cos nx \, dx$ and $\int_{-\pi}^{\pi} \sin nx \, dx$. As you know, the integral of a sine or cosine over a complete period is zero. This is obvious from a graph - if you are not already satisfied on this point a quick sketch should convince you. The period of cos nx is $2\pi/n$, and therefore in integrating from $-\pi$ to π we are integrating over n complete periods. Hence $\int_{-\pi}^{\pi} \cos nx \, dx = 0$.

What is the period of sin nx and what is the value of $\int_{-\pi}^{\pi} \sin nx \, dx$?

****

13A

The period of sin nx is $2\pi/n$.

$\int_{-\pi}^{\pi} \sin nx \, dx = 0$ *as the integral is over n complete periods.*

FRAME 14

The other integrals that will be needed are $\int_{-\pi}^{\pi} \cos mx \cos nx\, dx$, $\int_{-\pi}^{\pi} \sin mx \sin nx\, dx$ and $\int_{-\pi}^{\pi} \sin mx \cos nx\, dx$.

For the first one, we have

$$\int_{-\pi}^{\pi} \cos mx \cos nx\, dx = \tfrac{1}{2} \int_{-\pi}^{\pi} \{\cos(m+n)x + \cos(m-n)x\}\, dx$$

$$= 0 \quad \text{if } m \neq n, \text{ as both cosines are integrated over a number of complete periods.}$$

But if $m = n$, the integral becomes $\tfrac{1}{2} \int_{-\pi}^{\pi} (\cos 2nx + 1)\, dx$

$$= \pi$$

Now work out the other two integrals, not forgetting to consider whether $m = n$ must be treated as a special case.

****

14A

$$\int_{-\pi}^{\pi} \sin mx \sin nx\, dx = \tfrac{1}{2} \int_{-\pi}^{\pi} \{\cos(m-n)x - \cos(m+n)x\}\, dx$$

$$\begin{cases} = 0 & \text{if } m \neq n \\ = \pi & \text{if } m = n \end{cases}$$

$$\int_{-\pi}^{\pi} \sin mx \cos nx\, dx = \tfrac{1}{2} \int_{-\pi}^{\pi} \{\sin(m+n)x + \sin(m-n)x\}\, dx$$

$$= 0 \quad \text{in all cases.}$$

FRAME 15

The integrals worked out in the last two frames are now summarised for easy reference.

$$\int_{-\pi}^{\pi} \cos nx\, dx = 0 \qquad (15.1)$$

$$\int_{-\pi}^{\pi} \sin nx\, dx = 0 \qquad (15.2)$$

FOURIER SERIES

FRAME 15 continued

$$\int_{-\pi}^{\pi} \cos mx \cos nx \, dx = \begin{cases} 0 & m \neq n \\ \pi & m = n \end{cases} \quad \begin{array}{l}(15.3)\\(15.4)\end{array}$$

$$\int_{-\pi}^{\pi} \sin mx \sin nx \, dx = \begin{cases} 0 & m \neq n \\ \pi & m = n \end{cases} \quad \begin{array}{l}(15.5)\\(15.6)\end{array}$$

$$\int_{-\pi}^{\pi} \sin mx \cos nx \, dx = 0 \quad (15.7)$$

If we integrate these functions between other limits, we find that the values are the same so long as the limits differ by 2π

e.g.   0 to 2π, $\frac{-\pi}{2}$ to $\frac{3\pi}{2}$ etc.

FRAME 16

Calculation of the coefficients

We are now ready to find formulae for calculating the a's and b's.

First, we will show how to find a_o.

Now $f(x) = \tfrac{1}{2}a_o + a_1 \cos x + a_2 \cos 2x + \ldots + a_n \cos nx + \ldots$
$\qquad\qquad + b_1 \sin x + b_2 \sin 2x + \ldots + b_n \sin nx + \ldots$ (16.1)

and integrating both sides from $-\pi$ to π gives

$$\int_{-\pi}^{\pi} f(x) \, dx = \int_{-\pi}^{\pi} \tfrac{1}{2}a_o \, dx$$

The integrals of all the other terms will be zero, as each is a case of either (15.1) or (15.2).

We now get $\int_{-\pi}^{\pi} f(x) \, dx = \tfrac{1}{2}a_o \cdot 2\pi$

$$\therefore \; a_o = \frac{1}{\pi} \int_{-\pi}^{\pi} f(x) \, dx$$

Note that $\tfrac{1}{2}a_o = \frac{1}{2\pi} \int_{-\pi}^{\pi} f(x) \, dx$

$\qquad\qquad\qquad = $ mean value of $f(x)$ over the range $-\pi$ to π.

FRAME 17

The next problem is to find the coefficients of the cosine terms.
To find a_1, we shall multiply both sides of (16.1) by $\cos x$ and integrate from $-\pi$ to π. This gives

$$\int_{-\pi}^{\pi} f(x) \cos x \, dx = \int_{-\pi}^{\pi} a_1 \cos^2 x \, dx$$

Once again, the integrals of all the other terms will be zero, as each is a case of (15.1), (15.3) or (15.7). You should check this.

Now using (15.4) we have

$$\int_{-\pi}^{\pi} f(x) \cos x \, dx = a_1 \pi$$

$$\therefore a_1 = \frac{1}{\pi} \int_{-\pi}^{\pi} f(x) \cos x \, dx$$

Now, how do you suggest that we should find $a_2, a_3, \ldots, a_n$?

****

17A

To find a_2, a_3, etc. multiply both sides of (16.1) by $\cos 2x$, $\cos 3x$, etc. respectively and integrate between $-\pi$ and π.

In general, to find a_n multiply both sides of (16.1) by $\cos nx$ and integrate between $-\pi$ and π.

FRAME 18

Turning now to the case of the general term we suggest that you find the formula for a_n.

****

18A

$$\int_{-\pi}^{\pi} f(x) \cos nx \, dx = \int_{-\pi}^{\pi} a_n \cos^2 nx \, dx$$

$$= a_n \pi$$

$$\therefore a_n = \frac{1}{\pi} \int_{-\pi}^{\pi} f(x) \cos nx \, dx$$

FOURIER SERIES

FRAME 19

If we put $n = 2, 3, \ldots$ in the formula for a_n, we get

$$a_2 = \frac{1}{\pi} \int_{-\pi}^{\pi} f(x) \cos 2x \, dx$$

$$a_3 = \frac{1}{\pi} \int_{-\pi}^{\pi} f(x) \cos 3x \, dx$$

and so on. You will notice that putting $n = 1$ gives a_1, of course. Also, putting $n = 0$ gives the formula for a_0 obtained in FRAME 16 so that the general formula for a_n also covers this case. Now you will understand why we wrote c_0 as $\frac{1}{2}a_0$, rather than a_0, in (4.1).

We are now left with the problem of finding the coefficients of the sine terms i.e. the b's. Proceeding directly to the general case, can you suggest a method for finding the formula for b_n?

****

19A

To find b_n, multiply both sides of (16.1) by $\sin nx$ and integrate from $-\pi$ to π.

FRAME 20

By a method similar to those used previously, now find the formula for b_n.

****

20A

$$\int_{-\pi}^{\pi} f(x) \sin nx \, dx = \int_{-\pi}^{\pi} b_n \sin^2 nx \, dx.$$

$$= b_n \pi$$

$$\therefore b_n = \frac{1}{\pi} \int_{-\pi}^{\pi} f(x) \sin nx \, dx$$

FRAME 21

Summarising, we have the following result:

If a periodic function $f(x)$, of period 2π, can be represented by

$$f(x) = \tfrac{1}{2}a_0 + a_1 \cos x + a_2 \cos 2x + \ldots + a_n \cos nx + \ldots$$
$$\qquad\qquad + b_1 \sin x + b_2 \sin 2x + \ldots + b_n \sin nx + \ldots$$

then $a_n = \dfrac{1}{\pi} \displaystyle\int_{-\pi}^{\pi} f(x) \cos nx\, dx$

and $b_n = \dfrac{1}{\pi} \displaystyle\int_{-\pi}^{\pi} f(x) \sin nx\, dx.$

You will notice that a_n is twice the mean value of $f(x) \cos nx$ over a period and b_n is twice the mean value of $f(x) \sin nx$ over a period.

NOTE: As the integrals used in arriving at the formulae for a_n and b_n have the same values for any interval of 2π (see FRAME 15), we could use any such interval instead of $-\pi$ to π. Thus we could also have, for instance,

$$a_n = \frac{1}{\pi} \int_0^{2\pi} f(x) \cos nx\, dx$$

$$b_n = \frac{1}{\pi} \int_0^{2\pi} f(x) \sin nx\, dx$$

An alternative way of writing the series for $f(x)$, which is more compact, is to use the $\sum$ notation, i.e.

$$f(x) = \tfrac{1}{2}a_0 + \sum_{n=1}^{\infty} a_n \cos nx + \sum_{n=1}^{\infty} b_n \sin nx$$

FOURIER SERIES

FRAME 22

As a first example, let us consider the following waveform, which is of period 2π:

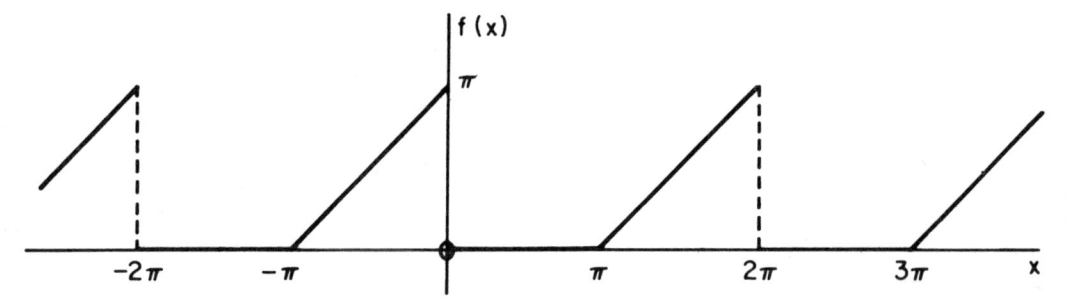

Can you give a set of equations which would define this function?

****

22A

Two possible sets are:

$$\left. \begin{array}{ll} f(x) = x + \pi & -\pi < x < 0 \\ f(x) = 0 & 0 < x < \pi \\ f(x + 2\pi) = f(x) & \end{array} \right\} \quad (22A.1)$$

$$\left. \begin{array}{ll} f(x) = 0 & 0 < x < \pi \\ f(x) = x - \pi & \pi < x < 2\pi \\ f(x + 2\pi) = f(x) & \end{array} \right\} \quad (22A.2)$$

FRAME 23

Using the set of equations (22A.1), the Fourier series for the wave-form would be found as follows:

$$a_o = \frac{1}{\pi} \int_{-\pi}^{\pi} f(x)\, dx$$

$$= \frac{1}{\pi} \left[\int_{-\pi}^{0} f(x)\, dx + \int_{0}^{\pi} f(x)\, dx \right]$$

The interval has to be split up in this way, because of the different f(x) in the two halves.

FRAME 23 continued

$$\therefore a_0 = \frac{1}{\pi}\left[\int_{-\pi}^{0}(x+\pi)dx + \int_{0}^{\pi} 0 \cdot dx\right]$$

The value of the second integral is clearly zero. The first one could be evaluated by ordinary integration, but we notice that it is the area of a triangle whose base is π and height is π.

$$\therefore a_0 = \frac{1}{\pi} \cdot \frac{\pi^2}{2} = \frac{\pi}{2}$$

$$a_n = \frac{1}{\pi}\int_{-\pi}^{\pi} f(x)\cos nx\, dx$$

$$= \frac{1}{\pi}\left[\int_{-\pi}^{0}(x+\pi)\cos nx\, dx + \int_{0}^{\pi} 0 \cdot \cos nx\, dx\right]$$

$$= \frac{1}{\pi}\int_{-\pi}^{0}(x+\pi)\cos nx\, dx + 0$$

$$= \frac{1}{\pi}\left\{\left[(x+\pi)\frac{1}{n}\sin nx\right]_{-\pi}^{0} - \int_{-\pi}^{0} 1 \cdot \frac{1}{n}\sin nx\, dx\right\} \quad \text{on integrating by parts}$$

$$= \frac{1}{\pi}\left\{0 - \left[-\frac{1}{n^2}\cos nx\right]_{-\pi}^{0}\right\}$$

$$= \frac{1}{\pi n^2}\left\{1 - \cos(-n\pi)\right\}$$

$$= \frac{1}{\pi n^2}(1 - \cos n\pi)$$

Now $\cos n\pi = 1$ if n is even, but -1 if n is odd.

$$\therefore a_n = \begin{cases} 0 & \text{if n is even} \\ \frac{2}{\pi n^2} & \text{if n is odd} \end{cases}$$

$$b_n = \frac{1}{\pi}\int_{-\pi}^{\pi} f(x)\sin nx\, dx \quad \text{and the evaluation of this is left to you.}$$

****

FOURIER SERIES

23A

$$b_n = \frac{1}{\pi} \int_{-\pi}^{0} (x + \pi) \sin nx \, dx$$

$$= \frac{1}{\pi} \left\{ \left[-(x + \pi)\frac{1}{n} \cos nx \right]_{-\pi}^{0} - \int_{-\pi}^{0} -\frac{1}{n} \cos nx \, dx \right\}$$

$$= \frac{1}{\pi} \left\{ \frac{-\pi}{n} + \frac{1}{n^2} \left[\sin nx \right]_{-\pi}^{0} \right\}$$

$$= -\frac{1}{n}$$

FRAME 24

The Fourier series for the waveform is, therefore,

$$\frac{\pi}{4} + \frac{2}{\pi}(\cos x + \frac{1}{3^2} \cos 3x + \frac{1}{5^2} \cos 5x + \ldots)$$

$$- (\sin x + \frac{1}{2} \sin 2x + \frac{1}{3} \sin 3x + \frac{1}{4} \sin 4x + \ldots)$$

Alternatively, using the $\sum$ notation, this could be written as

$$\frac{\pi}{4} + \frac{2}{\pi} \sum_{k=1}^{\infty} \frac{1}{(2k-1)^2} \cos(2k-1)x - \sum_{n=1}^{\infty} \frac{1}{n} \sin nx$$

We suggest that you now verify that the same values of a_o, a_n, and b_n, and hence the same series, are obtained if the set of equations (22A.2) is used.

****

24A

$$a_o = \frac{1}{\pi} \int_{0}^{2\pi} f(x) \, dx$$

$$= \frac{1}{\pi} \int_{\pi}^{2\pi} (x - \pi) \, dx$$

$$= \frac{\pi}{2}$$

$$a_n = \frac{1}{\pi} \int_{0}^{2\pi} f(x) \cos nx \, dx$$

24A continued

$$= \frac{1}{\pi} \int_{\pi}^{2\pi} (x - \pi) \cos nx \, dx$$

$$= \begin{cases} 0 & \text{if } n \text{ is even} \\ \frac{2}{\pi n^2} & \text{if } n \text{ is odd} \end{cases}$$

$$b_n = \frac{1}{\pi} \int_0^{2\pi} f(x) \sin nx \, dx$$

$$= \frac{1}{\pi} \int_\pi^{2\pi} (x - \pi) \sin nx \, dx$$

$$= -\frac{1}{n}$$

FRAME 25

In FRAME 5 we quoted the Fourier series for the square-wave function shown below.

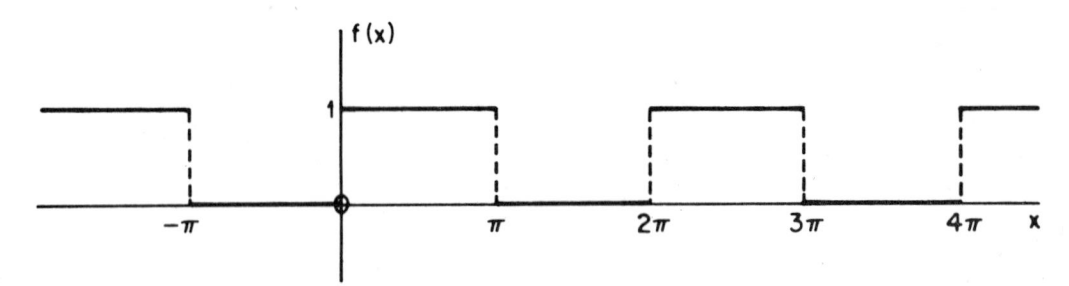

Now that we know how to calculate the coefficients, we can actually obtain the series.

Whether the interval for integration is taken as $-\pi$ to π or 0 to 2π, the formulae for the coefficients will become

$$a_o = \frac{1}{\pi} \int_0^\pi 1 \, dx = \frac{1}{\pi} \times \text{area of rectangle}$$

$$a_n = \frac{1}{\pi} \int_0^\pi 1 \cdot \cos nx \, dx$$

FOURIER SERIES

FRAME 25 continued

$$b_n = \frac{1}{\pi} \int_0^\pi 1 \cdot \sin nx \, dx$$

Now evaluate the coefficients and hence obtain the series.

****

25A

$$a_o = \frac{1}{\pi} \cdot \pi = 1$$

$$a_n = \frac{1}{\pi} \left[\frac{1}{n} \sin nx \right]_0^\pi = 0$$

$$b_n = \frac{1}{\pi} \left[-\frac{1}{n} \cos nx \right]_0^\pi$$

$$= \frac{1}{\pi n} (1 - \cos n\pi)$$

$$= \begin{cases} 0 & \text{if } n \text{ is even} \\ \frac{2}{\pi n} & \text{if } n \text{ is odd} \end{cases}$$

∴ The series is $\frac{1}{2} + \frac{2}{\pi}(\sin x + \frac{1}{3}\sin 3x + \frac{1}{5}\sin 5x + \frac{1}{7}\sin 7x + \ldots)$

FRAME 26

Odd and Even Functions

In the series just obtained some of the coefficients turned out to be zero. If we could tell in advance that this was going to happen, we could reduce the work done in calculating the coefficients.

It is particularly useful, in this context, to be able to recognise an odd or an even function. It may be helpful to remind you, at this stage, of the definition and properties of odd and even functions.

The function $f(x)$ is defined as being even if $f(-x) = f(x)$
"   "   "   "   "   "   "   odd " $f(-x) = -f(x)$

FRAME 26 continued

From the definition, any even power of x is seen to be an even function, likewise cos x and, more generally, cos nx. You will remember that the Maclaurin series for cos x (and other even functions) contains only even powers of x.

Similarly, any odd power of x is an odd function and so is also sin x, and, more generally, sin nx. The Maclaurin series for sin x (and other odd functions) contains only odd powers of x.

State which of the following are (a) even, (b) odd, (c) neither.

(i)   x cos x

(ii) x^2 cos 5x

(iii)   (x + π) cos x

(iv)   sin x sin 2x

****

26A

(i) $f(x) = x \cos x$

$f(-x) = (-x) \cos (-x)$

$ = -x \cos x$

$ = -f(x)$

∴ $x \cos x$ is an odd function

(ii)   Even

(iii)   Neither

(iv)   Even

You will notice that the product of two even functions, as in (ii), or of two odd functions, as in (iv), is even, and that the product of an even function and an odd function, as in (i), is odd.

FOURIER SERIES

FRAME 27

It follows from the definition of an even function that its graph is symmetrical about the y-axis. Similarly the graph of an odd function is symmetrical about the origin. This should be obvious to you if you consider the graphs of $y = x^2$ (an even function) and $y = x^3$ (an odd function).

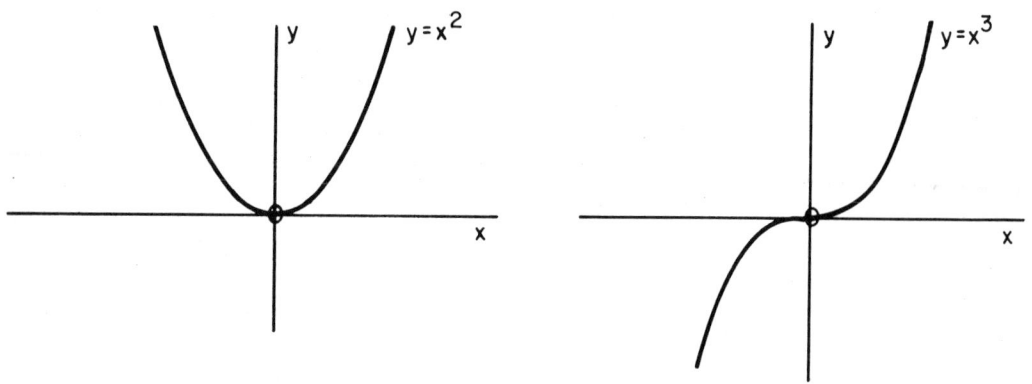

The graphs of $y = \cos x$ (an even function) and $y = \sin x$ (an odd function) show the same kinds of symmetry.

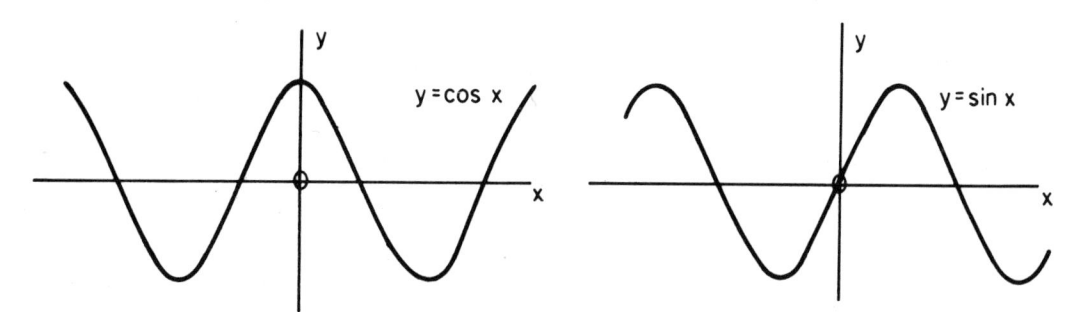

State which of the following graphs represent (a) an even function, (b) an odd function, (c) neither.

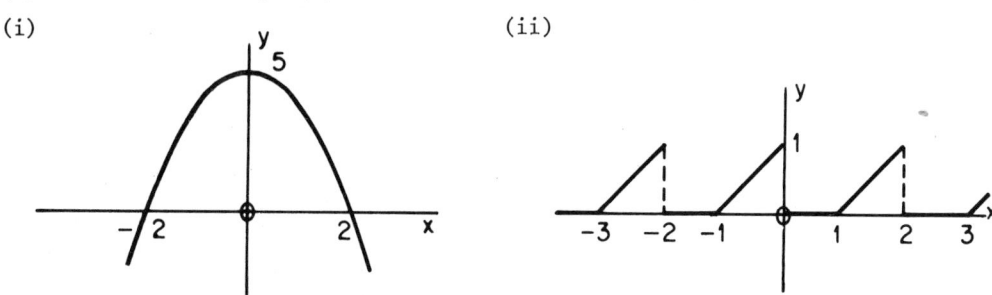

(i)

(ii)

FRAME 27 continued

(iii) [square wave graph, amplitude ±1, period 2π]

(iv) [sawtooth graph, amplitude ±π, period 2π]

(v) [horizontal line $y = a$]

****

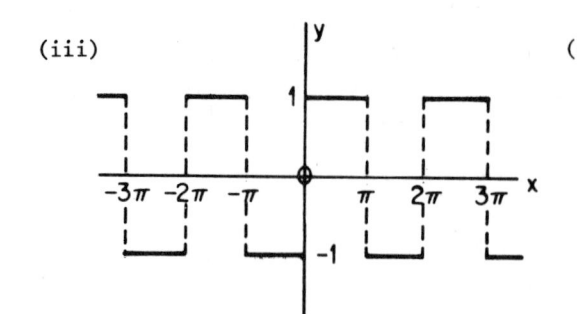

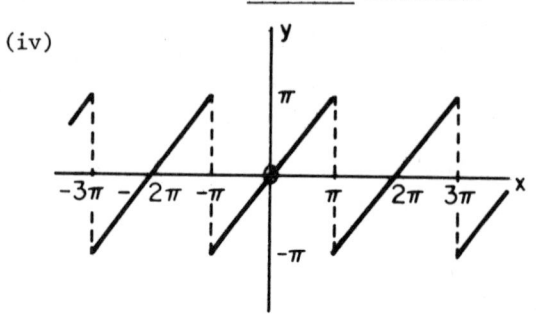

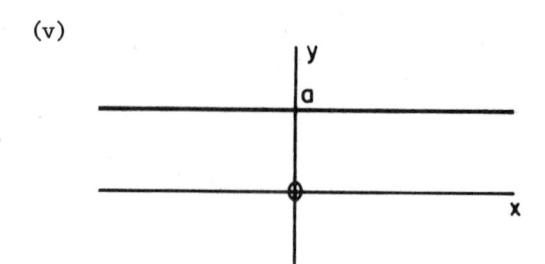

27A

(i)   Even   (ii) Neither
(iii) Odd   (iv) Odd
(v)   Even

FRAME 28

Integrals of the form $\int_{-h}^{h} f(x)\,dx$ occur in Fourier series (you have already met them with $h = \pi$) and it will be useful to find out if any conclusions can be drawn about the values of such integrals when $f(x)$ is known to be either an odd or even function. Looking at graphs (i) to (v) in the previous frame, can you see, in each case, any relationship between $\int_{0}^{h} y\,dx$ and $\int_{-h}^{h} y\,dx$ for the values of h given below?

For graphs (i)   and   (ii) $h = 1$
"   "   (iii) "   (iv) $h = \pi$
"   "   (v) $h = 5$

****

FOURIER SERIES

28A

(i) $\int_{-1}^{0} y\, dx = \int_{0}^{1} y\, dx$

(ii) No relationship

(iii) and (iv) $\int_{-\pi}^{0} y\, dx = -\int_{0}^{\pi} y\, dx$

(v) $\int_{-5}^{0} y\, dx = \int_{0}^{5} y\, dx$

FRAME 29

The following general conclusions will now be evident:

For any value of h, $\int_{-h}^{0} f(x)\, dx = \int_{0}^{h} f(x)\, dx$ if $f(x)$ is even, as in (i) and (v),

but $\int_{-h}^{0} f(x)\, dx = -\int_{0}^{h} f(x)\, dx$ if $f(x)$ is odd, as in (iii) and (iv).

What, therefore, can you deduce about the value of $\int_{-h}^{h} f(x)\, dx$ when $f(x)$ is (a) odd, (b) even?

****

29A

(a) When $f(x)$ is odd, $\int_{-h}^{h} f(x)\, dx = 0$

(b) When $f(x)$ is even, $\int_{-h}^{h} f(x)\, dx = 2\int_{0}^{h} f(x)\, dx$

(It is also $2\int_{-h}^{0} f(x)\, dx$ but negative limits are less convenient.)

FRAME 30

Obtaining the Fourier series for $f(x)$ is simplified in those cases where $f(x)$ can be seen to be either even or odd.

The Fourier series for an even function can only contain terms which are themselves even functions, i.e., it will have no sine terms and will be of the form

$$\tfrac{1}{2}a_0 + \sum_{n=1}^{\infty} a_n \cos nx.$$

On the other hand, the Fourier series for an odd function can only contain terms which are themselves odd functions, i.e., it will have only sine terms and will be of the form

$$\sum_{n=1}^{\infty} b_n \sin nx.$$

FRAME 31

Now, as an illustration, we shall consider the saw-tooth waveform shown in FRAME 27, graph (iv). It is described by the equations

$$f(x) = x \qquad -\pi < x < \pi$$
$$f(x + 2\pi) = f(x)$$

You have already noted that $f(x)$ is an odd function.

∴ In its Fourier series, $a_0 = a_n = 0$

$$b_n = \frac{1}{\pi} \int_{-\pi}^{\pi} x \sin nx \, dx$$

$$= \frac{2}{\pi} \int_{0}^{\pi} x \sin nx \, dx \quad \text{as} \quad x \sin nx \text{ is an even function}$$

$$= \frac{2}{\pi} \left\{ \left[-\frac{x}{n} \cos nx \right]_0^{\pi} - \int_0^{\pi} -\frac{1}{n} \cos nx \, dx \right\}$$

$$= \frac{2}{\pi} \left\{ -\frac{\pi}{n} \cos n\pi + \frac{1}{n} \left[\frac{1}{n} \sin nx \right]_0^{\pi} \right\}$$

$$= -\frac{2}{n} \cos n\pi$$

FOURIER SERIES 1:31

FRAME 31 continued

$$= \begin{cases} \dfrac{2}{n} & \text{if n is odd} \\ -\dfrac{2}{n} & \text{if n is even} \end{cases}$$

∴ The Fourier series for this waveform is

$$2(\sin x - \tfrac{1}{2}\sin 2x + \tfrac{1}{3}\sin 3x - \tfrac{1}{4}\sin 4x + \tfrac{1}{5}\sin 5x - \ldots)$$

Now obtain the Fourier series for the periodic function shown below.

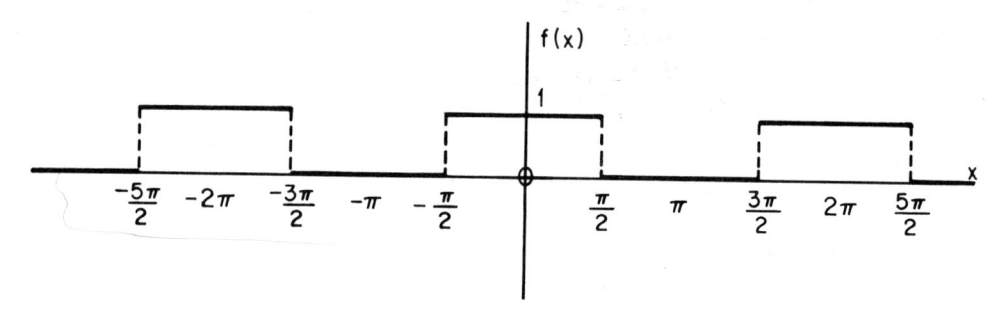

****

31A

This is an even function, so $b_n = 0$.

$\tfrac{1}{2}a_0$ = mean value of $f(x)$ over one period
     = $\tfrac{1}{2}$

$$a_n = \frac{1}{\pi}\int_{-\pi}^{\pi} f(x)\cos nx\, dx$$

$$= \frac{1}{\pi}\int_{-\pi/2}^{\pi/2} 1 \cdot \cos nx\, dx$$

$$= \frac{2}{\pi}\int_{0}^{\pi/2} \cos nx\, dx$$

$$= \frac{2}{\pi}\left[\frac{1}{n}\sin nx\right]_{0}^{\pi/2}$$

$$= \frac{2}{n\pi}\sin n\frac{\pi}{2}$$

31A continued

$$= \begin{cases} 0 & \text{if } n \text{ is even} \\ \dfrac{2}{n\pi} & \text{if } n \text{ is 1, 5, 9 etc.} \\ -\dfrac{2}{n\pi} & \text{if } n \text{ is 3, 7, 11 etc.} \end{cases}$$

∴ *The Fourier series is* $\dfrac{1}{2} + \dfrac{2}{\pi}(\cos x - \dfrac{1}{3}\cos 3x + \dfrac{1}{5}\cos 5x - \dfrac{1}{7}\cos 7x \ldots)$

FRAME 32

Sometimes, it may happen that a function which is neither odd nor even is symmetrical about a point on the y-axis, other than the origin. If this point is taken as a temporary origin, the function can then be treated as odd.

The square-wave function dealt with in FRAME 25 can be treated in this way. You will see that it is symmetrical about the point $(0, \tfrac{1}{2})$, and if a new origin is taken at this point, as shown in the diagram below, we shall then have an odd function $\phi(x)$.

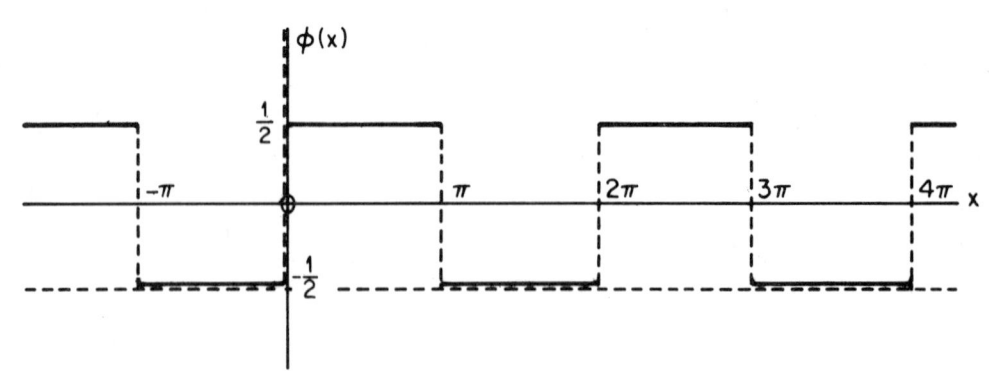

The original axes are shown as -------.

$\phi(x)$ is given by the equations
$$\begin{cases} \phi(x) = -\tfrac{1}{2} & -\pi < x < 0 \\ \phi(x) = \tfrac{1}{2} & 0 < x < \pi \\ \phi(x + 2\pi) = \phi(x) \end{cases}$$

Now find the Fourier series for $\phi(x)$.

****

32A

$\phi(x)$ is odd $\therefore$ $a_o = a_n = 0$

$$b_n = \frac{1}{\pi} \int_{-\pi}^{\pi} \phi(x) \sin nx \, dx$$

$$= \frac{2}{\pi} \int_0^{\pi} \tfrac{1}{2} \sin nx \, dx \quad \text{as} \quad \phi(x) \sin nx \text{ is an even function}$$

$$= \frac{1}{\pi} \left[-\frac{1}{n} \cos nx \right]_0^{\pi}$$

$$= \frac{1}{\pi n} (1 - \cos n\pi)$$

$$= \begin{cases} 0 & \text{if } n \text{ is even} \\ \frac{2}{\pi n} & \text{if } n \text{ is odd} \end{cases}$$

$\therefore \phi(x) = \frac{2}{\pi} \left(\sin x + \frac{1}{3} \sin 3x + \frac{1}{5} \sin 5x + \frac{1}{7} \sin 7x + \ldots \right)$

FRAME 33

Changing back to the original origin,

$$f(x) = \tfrac{1}{2} + \phi(x)$$
$$= \tfrac{1}{2} + \frac{2}{\pi}(\sin x + \tfrac{1}{3} \sin 3x + \tfrac{1}{5} \sin 5x + \tfrac{1}{7} \sin 7x + \ldots)$$

You will notice that this time we only had to work out b_n, and that the value of $\tfrac{1}{2}a_o$ is the distance of the new origin above the original one.

FRAME 34

Half-range Fourier Series

In FRAMES 8 – 11 we saw that some problems arise where the function is not periodic within the interval of definition but it is useful to represent it by a Fourier series. This difficulty was overcome by using a function which was periodic and which coincided with the given function over the interval of

FRAME 34 continued

definition. The period of the function we choose must obviously be greater than, or equal to, the given interval. Taking the interval as half a period makes it possible to obtain a simple series by defining the function for the other half period in such a way that it is either odd, thus giving a series of sines only, or even, giving a series of cosines only (including the term $\frac{1}{2}a_o$ as this can be regarded as $\frac{1}{2}a_o \cos 0x$).

For example, if we have the function defined between 0 and π as shown in the diagram

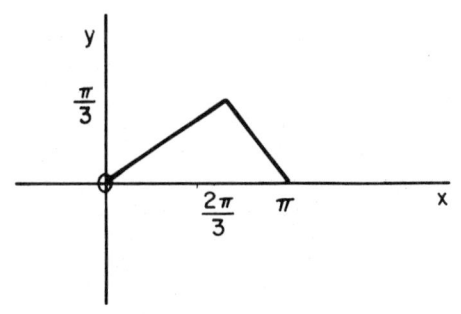

either of the following waveforms can be chosen to represent it.

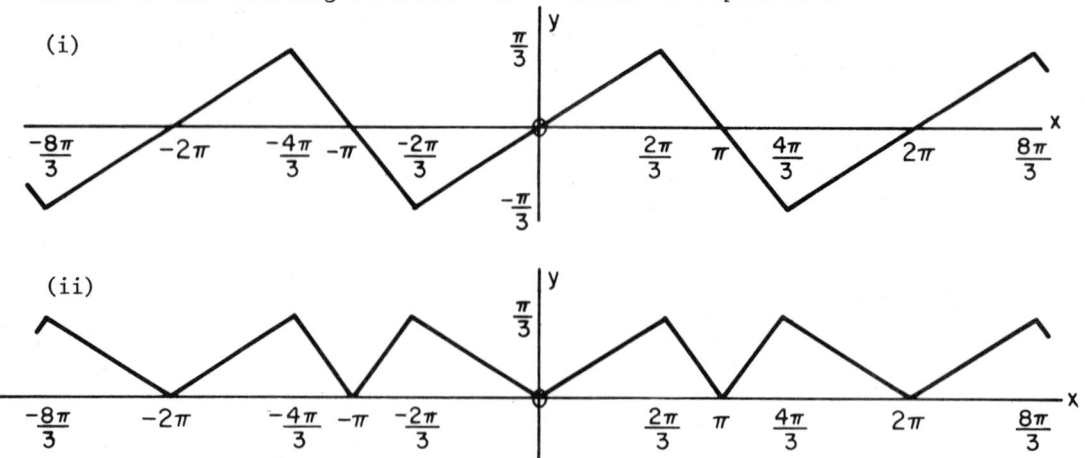

(i)

(ii)

(i) is an odd waveform and its series will have sines only (called a half-range Fourier sine series).

(ii) is an even waveform and its series will have cosines only (called a half-range Fourier cosine series).

FOURIER SERIES

FRAME 35

In half-range Fourier series, the calculation of the coefficients is simplified by the assumption of either an odd or an even function to represent the required function over the given interval.

If we assume an odd function $f(x)$, then $a_o = a_n = 0$

and $b_n = \dfrac{1}{\pi} \displaystyle\int_{-\pi}^{\pi} f(x) \sin nx\, dx$

$\phantom{\text{and } b_n} = \dfrac{2}{\pi} \displaystyle\int_{0}^{\pi} f(x) \sin nx\, dx \quad$ as $\quad f(x) \sin nx \quad$ is even.

If we assume an even function $f(x)$, then

$$b_n = 0$$

$$a_o = \dfrac{1}{\pi} \int_{-\pi}^{\pi} f(x)\, dx = \dfrac{2}{\pi} \int_{0}^{\pi} f(x)\, dx$$

and $\quad a_n = \dfrac{1}{\pi} \displaystyle\int_{-\pi}^{\pi} f(x) \cos nx\, dx$

$\phantom{\text{and } a_n} = \dfrac{2}{\pi} \displaystyle\int_{0}^{\pi} f(x) \cos nx\, dx \quad$ as $\quad f(x) \cos nx$ is even.

FRAME 36

Turning now to the example we used in FRAME 34, let us calculate the series for Case (i).

Between 0 and π the function is defined by the equations

$$\begin{cases} f(x) = \tfrac{1}{2}x & 0 < x < \dfrac{2\pi}{3} \\ f(x) = \pi - x & \dfrac{2\pi}{3} < x < \pi \end{cases}$$

$b_n = \dfrac{2}{\pi} \displaystyle\int_{0}^{\pi} f(x) \sin nx\, dx$

FRAME 36 continued

$$= \frac{2}{\pi}\left\{\int_0^{2\pi/3} \tfrac{1}{2}x \sin nx\, dx + \int_{2\pi/3}^{\pi} (\pi - x)\sin nx\, dx\right\}$$

Now show that this becomes $\dfrac{3}{\pi n^2}\sin\dfrac{2n\pi}{3}$ and then write down the first few terms of the series.

****

36A

$$b_n = \frac{2}{\pi}\left\{\tfrac{1}{2}\left[-\tfrac{x}{n}\cos nx\right]_0^{2\pi/3} - \tfrac{1}{2}\int_0^{2\pi/3}-\tfrac{1}{n}\cos nx\, dx + \left[-\tfrac{\pi-x}{n}\cos nx\right]_{2\pi/3}^{\pi}\right.$$
$$\left. -\int_{2\pi/3}^{\pi}\tfrac{1}{n}\cos nx\, dx\right\}$$

$$= \frac{2}{\pi}\left\{-\tfrac{\pi}{3n}\cos\tfrac{2n\pi}{3} + \tfrac{1}{2n^2}\sin\tfrac{2n\pi}{3} + \tfrac{\pi}{3n}\cos\tfrac{2n\pi}{3} + \tfrac{1}{n^2}\sin\tfrac{2n\pi}{3}\right\}$$

$$= \frac{3}{\pi n^2}\sin\frac{2n\pi}{3}$$

$$= \begin{cases} 3\sqrt{3}/2\pi n^2 & \text{if } n = 1, 4, 7, \ldots \\ -3\sqrt{3}/2\pi n^2 & \text{if } n = 2, 5, 8, \ldots \\ 0 & \text{if } n = 3, 6, 9, \ldots \end{cases}$$

$$\therefore\quad f(x) = \frac{3\sqrt{3}}{2\pi}\left(\sin x - \tfrac{1}{2^2}\sin 2x + \tfrac{1}{4^2}\sin 4x - \tfrac{1}{5^2}\sin 5x + \ldots\right)$$

FRAME 37

A function $f(x)$ is defined between 0 and π by the equation $f(x) = \pi - x$. Sketch the waveforms which will result if this is represented by a half-range Fourier series involving

        (i)   sines,
       (ii)   cosines,

and obtain the series in each case.

****

FOURIER SERIES

37A

(i)

$b_n = \frac{2}{\pi} \int_0^\pi (\pi - x) \sin nx \, dx = \frac{2}{n}$

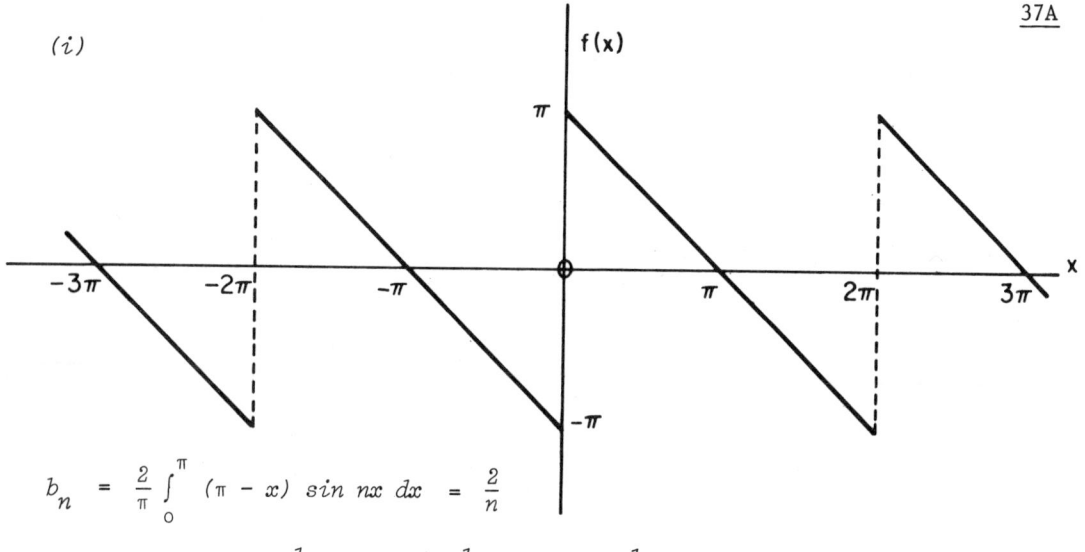

$f(x) = 2(\sin x + \frac{1}{2} \sin 2x + \frac{1}{3} \sin 3x + \frac{1}{4} \sin 4x + \ldots)$

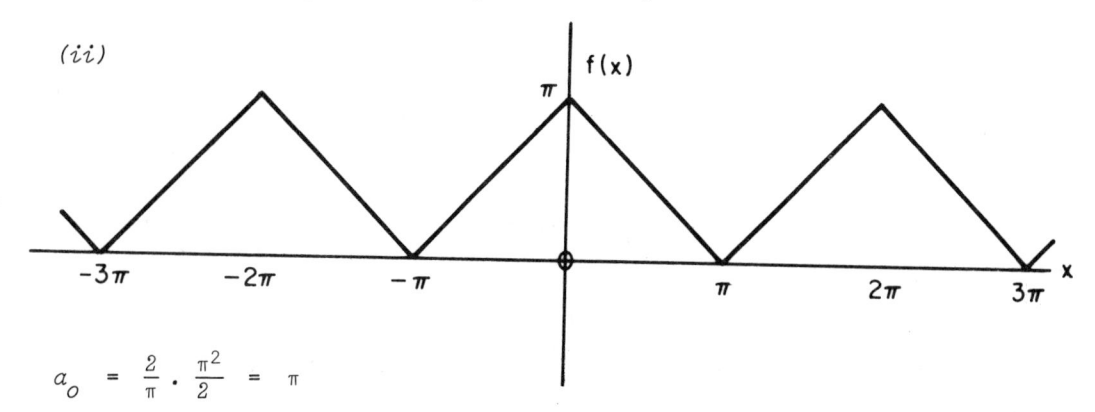

(ii)

$a_o = \frac{2}{\pi} \cdot \frac{\pi^2}{2} = \pi$

$a_n = \frac{2}{\pi} \int_0^\pi (\pi - x) \cos nx \, dx$

$ = \frac{2}{\pi n^2} (1 - \cos n\pi)$

$ = \begin{cases} 0 & \text{if } n \text{ is even} \\ \frac{4}{\pi n^2} & \text{if } n \text{ is odd} \end{cases}$

$f(x) = \frac{\pi}{2} + \frac{4}{\pi}(\cos x + \frac{1}{9} \cos 3x + \frac{1}{25} \cos 5x + \ldots)$

FRAME 38

Odd and Even Harmonics

We have seen that, if a function is odd or even, it is helpful to notice this from the graph before starting to calculate the coefficients. Other special types of waveform which may occur are those whose Fourier series contain only even harmonics or only odd harmonics. Again, it is useful to be able to see from the graph that this is going to happen.

If $f(x) = \frac{1}{2}a_0 + a_1 \cos x + a_2 \cos 2x + a_3 \cos 3x + \ldots$
$\qquad + b_1 \sin x + b_2 \sin 2x + b_3 \sin 3x + \ldots$

then, replacing x by $x + \pi$ will give

$f(x + \pi) = \frac{1}{2}a_0 + a_1 \cos(x + \pi) + a_2 \cos(2x + 2\pi) + a_3 \cos(3x + 3\pi) + \ldots$
$\qquad + b_1 \sin(x + \pi) + b_2 \sin(2x + 2\pi) + b_3 \sin(3x + 3\pi) + \ldots$

$\qquad = \frac{1}{2}a_0 - a_1 \cos x + a_2 \cos 2x - a_3 \cos 3x + \ldots$
$\qquad\qquad - b_1 \sin x + b_2 \sin 2x - b_3 \sin 3x + \ldots$

If the series for $f(x)$ contains even harmonics only ($\frac{1}{2}a_0$ is included as it may be considered as $\frac{1}{2}a_0 \cos 0x$), it can be seen that

$$f(x + \pi) = f(x)$$

On the other hand, if it contains odd harmonics only then

$$f(x + \pi) = -f(x)$$

An example of each case is shown below.

EVEN HARMONICS ONLY

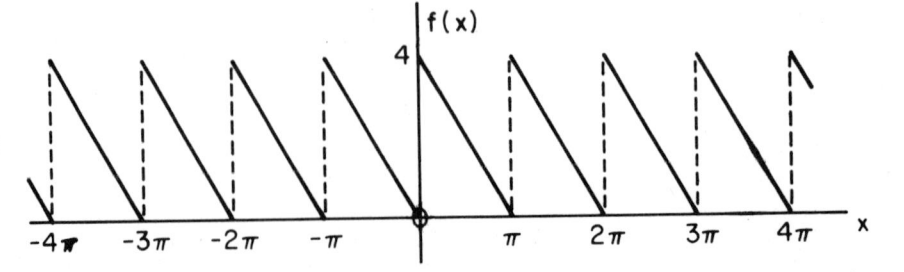

FOURIER SERIES

FRAME 38 continued

ODD HARMONICS ONLY

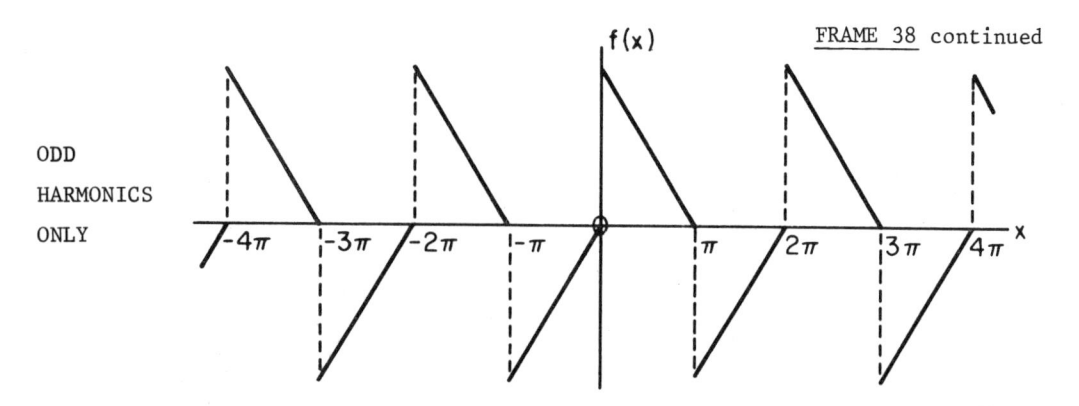

You will probably have realised that, in the first case, the function, as well as repeating itself at intervals of 2π, also repeats itself at intervals of π, as is implied by the relation $f(x + \pi) = f(x)$. This could also be taken as an example of the more general case where the period is other than 2π, which we shall shortly be considering.

Now state whether the series representing the following functions will contain (a) only even harmonics, (b) only odd harmonics, (c) both even and odd harmonics.

(i)

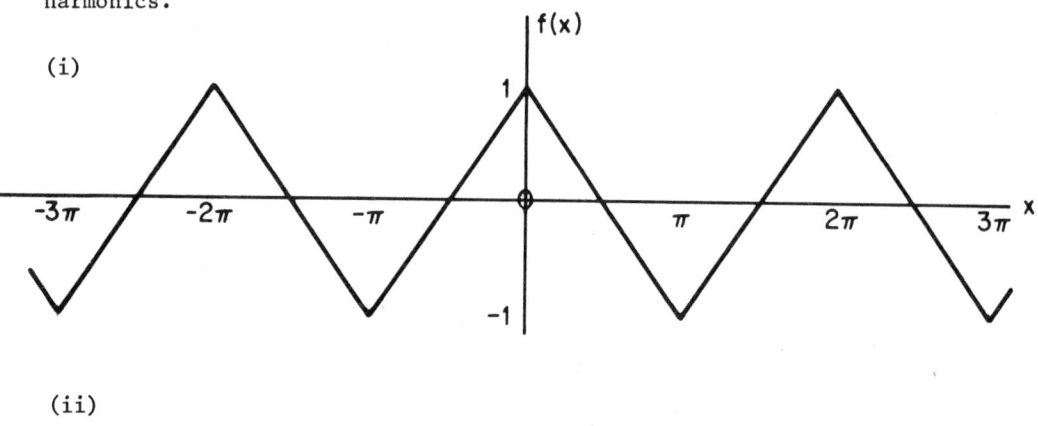

(ii)

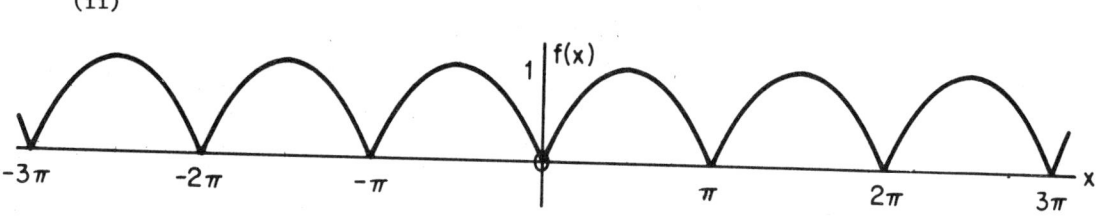

.c

FRAME 38 continued

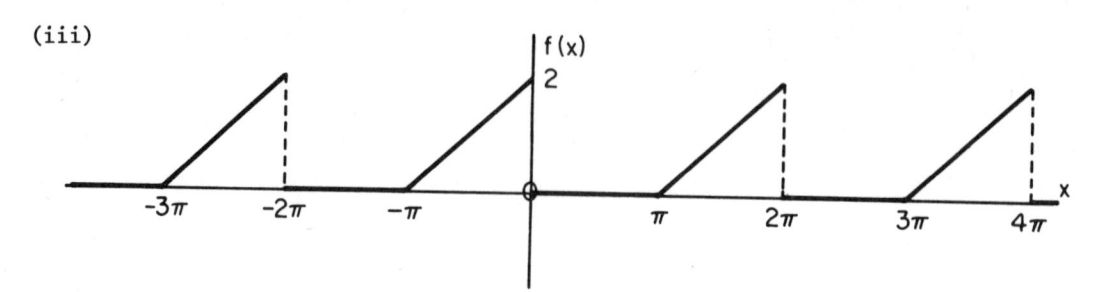

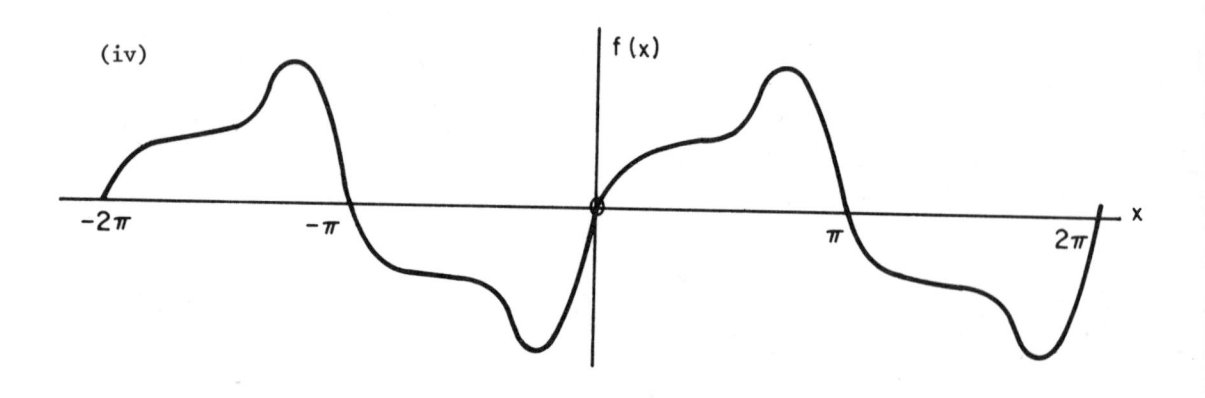

(iii)

[Graph showing sawtooth-like waveform with peaks of 2 at x = -2π, 0, 2π, 4π]

(iv)

[Graph showing odd-symmetric waveform across -2π to 2π]

****

38A

(i)   Only odd harmonics.

(ii) Only even harmonics.   (The rectified sine wave.)

(iii) Both.

(iv) Only odd harmonics.

A voltage generated by a rotating machine is an example of a waveform in engineering whose Fourier series contains only odd harmonics. As a result of the similarity between the N and S magnetic poles in any such machine, the voltage generated always has this type of symmetry.

FOURIER SERIES

FRAME 39

Sum of a Fourier series at a point of discontinuity

If you substitute a particular value of x into a Fourier series, the sum which results is the corresponding value of f(x).

If, however, this is done at a point of discontinuity x = k, say, the sum is the average of the two limiting values of f(x) as x → k from the left and right respectively.

The series for the waveform

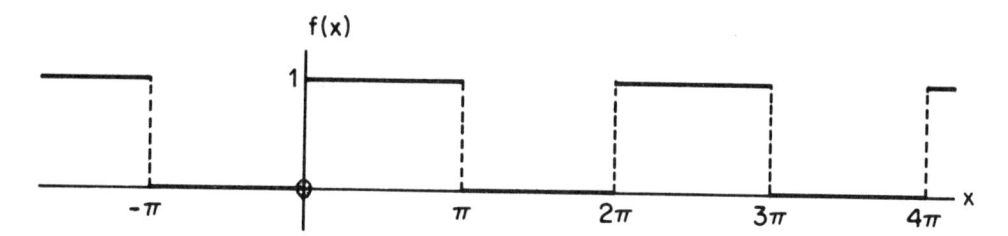

is $\frac{1}{2} + \frac{2}{\pi} \left(\sin x + \frac{1}{3} \sin 3x + \frac{1}{5} \sin 5x + \frac{1}{7} \sin 7x + \ldots \right)$ (See FRAME 25)

Substituting x = π, for instance, in this series will give the value $\frac{1}{2}$ which is $\frac{1}{2}(1 + 0)$.

FRAME 40

General Period

So far we have only considered functions of period 2π, but as we saw in FRAMES 6 - 11, most practical problems involve the use of functions with a different period. We shall now look at an example of a function whose period is not 2π - the waveform is shown below.

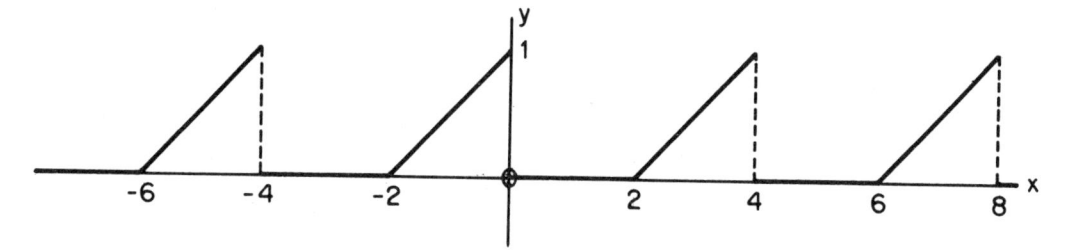

FRAME 40 continued

The equation of this is y = f(x) where

$$\begin{cases} f(x) = \tfrac{1}{2}(x+2) & -2 < x < 0 \\ f(x) = 0 & 0 < x < 2 \\ f(x+4) = f(x) & \text{i.e. the period is 4.} \end{cases}$$

This can, in fact, be converted to the straightforward case where the period is 2π by a change in the horizontal scale as illustrated below.

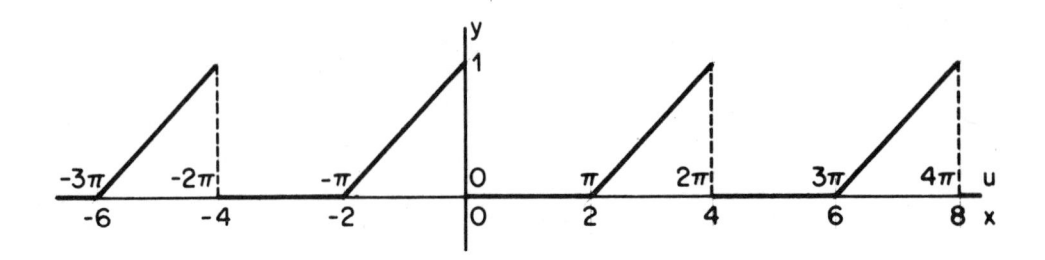

We now treat u as a new independent variable, and hence the equation of the waveform can be written in the form y = F(u).

From the graph write down:

(i)   The equations which define F(u). (Ignore the x-scale while you are doing this.)

(ii)   The equation which gives u in terms of x.

****

40A

$$\begin{cases} F(u) = \tfrac{1}{\pi}(u+\pi) & -\pi < u < 0 \\ F(u) = 0 & 0 < u < \pi \\ F(u+2\pi) = F(u) \end{cases}$$

$$u = \frac{\pi x}{2}$$

FOURIER SERIES

FRAME 41

We now have a function $F(u)$ which has period 2π. Show that the Fourier series for this is

$$F(u) = \frac{1}{4} + \frac{2}{\pi^2}\left(\cos u + \frac{1}{3^2}\cos 3u + \frac{1}{5^2}\cos 5u + \ldots\right)$$

$$- \frac{1}{\pi}\left(\sin u + \frac{1}{2}\sin 2u + \frac{1}{3}\sin 3u + \ldots\right)$$

****

41A

$$a_o = \frac{1}{\pi} \cdot \frac{\pi}{2} = \frac{1}{2}$$

$$a_n = \frac{1}{\pi}\int_{-\pi}^{\pi} F(u) \cos nu \, du$$

$$= \frac{1}{\pi}\int_{-\pi}^{0} \frac{1}{\pi}(u + \pi) \cos nu \, du$$

$$= \frac{1}{\pi^2 n^2}(1 - \cos n\pi)$$

$$= \begin{cases} 0 & \text{if } n \text{ is even} \\ \frac{2}{\pi^2 n^2} & \text{if } n \text{ is odd} \end{cases}$$

$$b_n = \frac{1}{\pi}\int_{-\pi}^{\pi} F(u) \sin nu \, du$$

$$= \frac{1}{\pi}\int_{-\pi}^{0} \frac{1}{\pi}(u + \pi) \sin nu \, du$$

$$= -\frac{1}{\pi n}$$

$$\therefore F(u) = \frac{1}{4} + \frac{2}{\pi^2}\left(\cos u + \frac{1}{3^2}\cos 3u + \frac{1}{5^2}\cos 5u + \ldots\right)$$

$$- \frac{1}{\pi}\left(\sin u + \frac{1}{2}\sin 2u + \frac{1}{3}\sin 3u + \ldots\right)$$

FRAME 42

Now $F(u) = y = f(x)$ and $u = \frac{\pi x}{2}$, so the Fourier series for $f(x)$ is

$$f(x) = \frac{1}{4} + \frac{2}{\pi^2}\left(\cos\frac{\pi x}{2} + \frac{1}{3^2}\cos\frac{3\pi x}{2} + \frac{1}{5^2}\cos\frac{5\pi x}{2} + \ldots\right)$$

$$- \frac{1}{\pi}\left(\sin\frac{\pi x}{2} + \frac{1}{2}\sin \pi x + \frac{1}{3}\sin\frac{3\pi x}{2} + \ldots\right)$$

Having seen how to deal with an example of this type, in the next frame we shall consider the general case with period 2ℓ, where ℓ can take any value.

FRAME 43

When the period is 2ℓ, we again take a new horizontal scale, but this time chosen so that $u = 2\pi$ corresponds to $x = 2\ell$.

-3π	-2π	$-\pi$	0	π	2π	3π	u
-3ℓ	-2ℓ	$-\ell$	0	ℓ	2ℓ	3ℓ	x

Write down the equation giving u in terms of x.

****

43A

$$u = \frac{\pi x}{\ell}$$

FRAME 44

$y = f(x)$ now becomes $y = F(u)$ with period 2π, and the Fourier series for $F(u)$ will be

$$\tfrac{1}{2}a_0 + a_1 \cos u + a_2 \cos 2u + a_3 \cos 3u + \ldots$$
$$+ b_1 \sin u + b_2 \sin 2u + b_3 \sin 3u + \ldots$$

where $a_0 = \frac{1}{\pi}\int_{-\pi}^{\pi} F(u)\,du$

$a_n = \frac{1}{\pi}\int_{-\pi}^{\pi} F(u) \cos nu\,du$

FRAME 44 continued

$$b_n = \frac{1}{\pi} \int_{-\pi}^{\pi} F(u) \sin nu \, du$$

The series and formulae for the coefficients can be written in terms of x, using $u = \frac{\pi x}{\ell}$, to give the Fourier series for $f(x)$.

Thus $f(x) = \frac{1}{2}a_0 + a_1 \cos \frac{\pi x}{\ell} + a_2 \cos \frac{2\pi x}{\ell} + a_3 \cos \frac{3\pi x}{\ell} + \ldots$

$$+ b_1 \sin \frac{\pi x}{\ell} + b_2 \sin \frac{2\pi x}{\ell} + b_3 \sin \frac{3\pi x}{\ell} + \ldots$$

where $a_0 = \frac{1}{\pi} \int_{-\ell}^{\ell} f(x) \frac{\pi}{\ell} \, dx$ as $du = \frac{\pi}{\ell} dx$ and $x = \pm \ell$ corresponds to $u = \pm \pi$

$$= \frac{1}{\ell} \int_{-\ell}^{\ell} f(x) \, dx$$

$$a_n = \frac{1}{\ell} \int_{-\ell}^{\ell} f(x) \cos \frac{n\pi x}{\ell} \, dx$$

$$b_n = \frac{1}{\ell} \int_{-\ell}^{\ell} f(x) \sin \frac{n\pi x}{\ell} \, dx$$

These can be regarded as standard formulae for the coefficients for the general period, so that it is not necessary to make the u substitution every time.

You will recall that in FRAME 21 we pointed out that for a function of period 2π, the integrals used in evaluating the coefficients can be taken over <u>any</u> interval of 2π.

Similarly, here the integrals can be taken over <u>any</u> interval of 2ℓ. For example, we could use limits 0 and 2ℓ instead of $-\ell$ and ℓ.

FRAME 45

As an example, consider the waveform shown below.

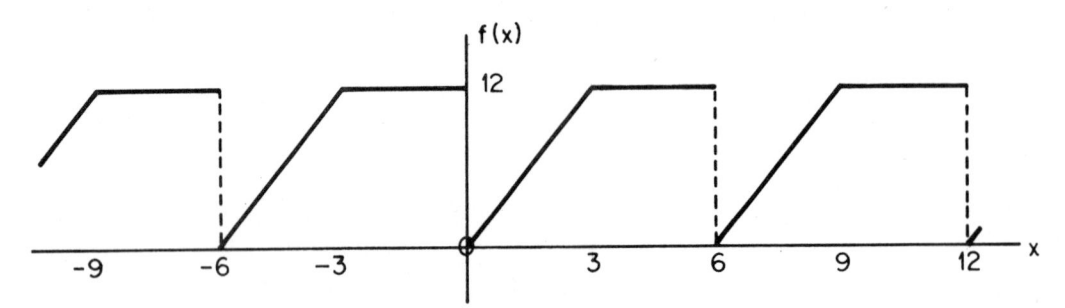

First, write down the equations defining the function and state the value of ℓ in this case.

****

45A

$$\begin{cases} f(x) = 12 & -3 < x < 0 \\ f(x) = 4x & 0 < x < 3 \\ f(x+6) = f(x) \end{cases} \quad \text{OR} \quad \begin{cases} f(x) = 4x & 0 < x < 3 \\ f(x) = 12 & 3 < x < 6 \\ f(x+6) = f(x) \end{cases}$$

The period is 6, so the value of ℓ is 3.

FRAME 46

Now write down, but do not evaluate, the formulae for a_o, a_n and b_n for this particular waveform, using the standard results obtained in FRAME 43.

****

46A

$$a_o = \frac{1}{3}\left\{\int_{-3}^{0} 12\, dx + \int_{0}^{3} 4x\, dx\right\} \quad (= \frac{1}{3} \times \text{area under graph between } -3 \text{ and } +3)$$

$$a_n = \frac{1}{3}\left\{\int_{-3}^{0} 12 \cos \frac{n\pi x}{3}\, dx + \int_{0}^{3} 4x \cos \frac{n\pi x}{3}\, dx\right\}$$

$$b_n = \frac{1}{3}\left\{\int_{-3}^{0} 12 \sin \frac{n\pi x}{3}\, dx + \int_{0}^{3} 4x \sin \frac{n\pi x}{3}\, dx\right\}$$

FOURIER SERIES

46A continued

OR

$$a_o = \frac{1}{3}\left\{\int_0^3 4x\, dx + \int_3^6 12\, dx\right\} \quad (= \frac{1}{3} \times \text{area under graph between 0 and 6})$$

$$a_n = \frac{1}{3}\left\{\int_0^3 4x \cos\frac{n\pi x}{3}\, dx + \int_3^6 12 \cos\frac{n\pi x}{3}\, dx\right\}$$

$$b_n = \frac{1}{3}\left\{\int_0^3 4x \sin\frac{n\pi x}{3}\, dx + \int_3^6 12 \sin\frac{n\pi x}{3}\, dx\right\}$$

FRAME 47

Proceeding to the evaluation of the coefficients,

$$a_o = \frac{1}{3} \times 54 = 18$$

$$a_n = \frac{4}{3}\left\{\int_{-3}^0 3 \cos\frac{n\pi x}{3}\, dx + \int_0^3 x \cos\frac{n\pi x}{3}\, dx\right\}$$

$$= \frac{4}{3}\left\{\frac{9}{n\pi}\left[\sin\frac{n\pi x}{3}\right]_{-3}^0 + \left[x \cdot \frac{3}{n\pi}\sin\frac{n\pi x}{3}\right]_0^3 - \int_0^3 \frac{3}{n\pi}\sin\frac{n\pi x}{3}\, dx\right\}$$

$$= \frac{4}{n\pi} \cdot \frac{3}{n\pi}\left[\cos\frac{n\pi x}{3}\right]_0^3$$

$$= \frac{12}{n^2\pi^2}(\cos n\pi - 1)$$

$$= \begin{cases} 0 & \text{if } n \text{ is even} \\ -\frac{24}{n^2\pi^2} & \text{if } n \text{ is odd} \end{cases}$$

Now show that $b_n = -\frac{12}{n\pi}$ and write down the series for the waveform.

****

47A

$$9 - \frac{24}{\pi^2}\left(\cos\frac{\pi x}{3} + \frac{1}{3^2}\cos \pi x + \frac{1}{5^2}\cos\frac{5\pi x}{3} + \ldots\right)$$

$$- \frac{12}{\pi}\left(\sin\frac{\pi x}{3} + \frac{1}{2}\sin\frac{2\pi x}{3} + \frac{1}{3}\sin \pi x + \ldots\right)$$

FRAME 48

You will now easily see that the idea of a half-range Fourier series can be extended to a half-range other than π. This will entail the modification of the results in FRAME 44 for period 2ℓ, as in FRAME 35 for period 2π.

Thus, for an odd function, $a_o = a_n = 0$

and $b_n = \dfrac{2}{\ell} \displaystyle\int_0^\ell f(x) \sin \dfrac{n\pi x}{\ell} dx$

Now write down the corresponding results for an even function.

****

48A

$b_n = 0$

$a_o = \dfrac{2}{\ell} \displaystyle\int_0^\ell f(x) \, dx$

$a_n = \dfrac{2}{\ell} \displaystyle\int_0^\ell f(x) \cos \dfrac{n\pi x}{\ell} \, dx$

FRAME 49

Now try the following example.

Find the half-range Fourier sine series for x^2 in the interval $0 < x < 3$, and sketch the waveform represented by this series over several periods.

****

49A

$b_n = \dfrac{2}{3} \displaystyle\int_0^3 x^2 \sin \dfrac{n\pi x}{3} \, dx$

$= \dfrac{2}{3} \left\{ \left[-\dfrac{3x^2}{n\pi} \cos \dfrac{n\pi x}{3} \right]_0^3 - \displaystyle\int_0^3 2x \left(\dfrac{-3}{n\pi}\right) \cos \dfrac{n\pi x}{3} \, dx \right\}$

$= \dfrac{2}{3} \left\{ -\dfrac{27}{n\pi} \cos n\pi + \dfrac{6}{n\pi} \left[\dfrac{3x}{n\pi} \sin \dfrac{n\pi x}{3} \right]_0^3 - \dfrac{6}{n\pi} \displaystyle\int_0^3 \dfrac{3}{n\pi} \sin \dfrac{n\pi x}{3} \, dx \right\}$

$= \dfrac{2}{3} \left\{ -\dfrac{27}{n\pi} \cos n\pi - \dfrac{18}{n^2\pi^2} \left[-\dfrac{3}{n\pi} \cos \dfrac{n\pi x}{3} \right]_0^3 \right\}$

FOURIER SERIES

49A continued

$$= \frac{2}{3}\left\{-\frac{27}{n\pi}\cos n\pi + \frac{54}{n^3\pi^3}(\cos n\pi - 1)\right\}$$

$$= \begin{cases} -\dfrac{18}{n\pi} & \text{if } n \text{ is even} \\[2ex] \dfrac{18}{n\pi} - \dfrac{72}{n^3\pi^3} & \text{if } n \text{ is odd} \end{cases}$$

The series is $18\left\{\left(\dfrac{1}{\pi} - \dfrac{4}{\pi^3}\right)\sin\dfrac{\pi x}{3} - \dfrac{1}{2\pi}\sin\dfrac{2\pi x}{3} + \left(\dfrac{1}{3\pi} - \dfrac{4}{3^3\pi^3}\right)\sin \pi x$

$\qquad\qquad\qquad - \dfrac{1}{4\pi}\sin\dfrac{4\pi x}{3} + \ldots \right\}$

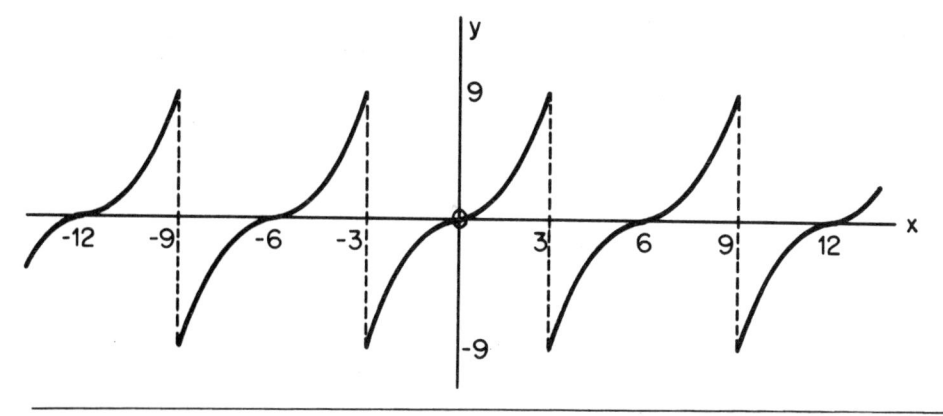

FRAME 50

Numerical Harmonic Analysis

The formulae obtained for calculating the coefficients in previous frames involved integrals whose evaluation depended on knowing the equation for y in terms of x. However, this may not be known and in its place we are only given or can measure numerical values of y for certain values of x. In such cases it is necessary to modify the formulae for the coefficients so that a numerical method can be used for their evaluation.

As an example:

If an alternating voltage is applied to a circuit consisting of a rectifier, a resistance and an inductance in series, the current is of the form

FRAME 50 continued

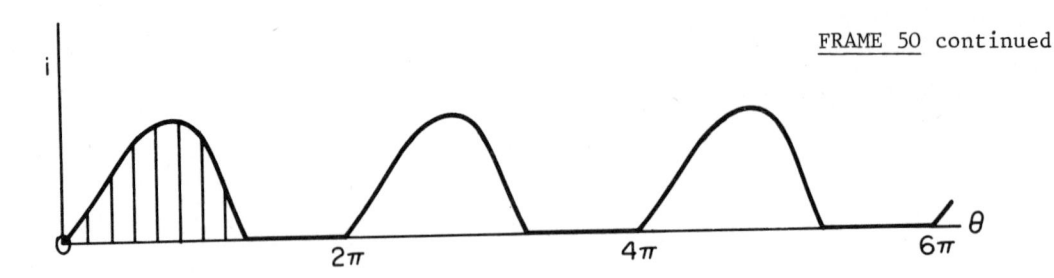

In a particular case the values of i, measured at 12 equidistant values of θ covering a period of 2π, were:

θ	0	π/6	π/3	π/2	2π/3	5π/6	π	7π/6	4π/3	3π/2	5π/3	11π/6
i	0	2·3	5·5	8·9	10·8	11·4	9·9	4·8	0	0	0	0

The coefficients in the Fourier series for $i = f(\theta)$ are found as follows:

$$a_o = \frac{1}{\pi} \int_0^{2\pi} f(\theta)\, d\theta \qquad = 2 \times \text{mean value of } f(\theta) \text{ over one period}$$

$$a_n = \frac{1}{\pi} \int_0^{2\pi} f(\theta) \cos n\theta\, d\theta = 2 \times \text{mean value of } f(\theta) \cos n\theta \text{ over one period}$$

$$b_n = \frac{1}{\pi} \int_0^{2\pi} f(\theta) \sin n\theta\, d\theta = 2 \times \text{mean value of } f(\theta) \sin n\theta \text{ over one period}$$

$f(\theta)$ is not known, so an approximation to the mean value of $f(\theta)$ is found by taking the mean of the 12 values of i at the tabulated θ's,

i.e. $\dfrac{\Sigma i}{12}$

Similarly the mean value of $f(\theta) \cos n\theta$ is taken as

$$\frac{\Sigma\, i \cos n\theta}{12}$$

and that of $f(\theta) \sin n\theta$ as

$$\frac{\Sigma\, i \sin n\theta}{12},$$

Thus $a_o = 2 \times \dfrac{\Sigma i}{12} = \dfrac{\Sigma i}{6}$, $\quad a_1 = \dfrac{\Sigma\, i \cos \theta}{6}$, $\quad b_1 = \dfrac{\Sigma\, i \sin \theta}{6}$ etc.

The calculation is best set out in a tabular form. For convenience, we shall change to θ in degrees.

FOURIER SERIES

FRAME 50 continued

$\theta°$	i	cos θ	i cos θ	sin θ	i sin θ	cos 2θ	i × cos 2θ	sin 2θ	i × sin 2θ
0	0	1	0	0	0	1	0	0	0
30	2·3	0·866	1·99	0·5	1·15	0·5	1·15	0·866	1·99
60	5·5	0·5	2·75	0·866	4·76	−0·5	−2·75	0·866	4·76
90	8·9	0	0	1	8·9	−1	−8·9	0	0
120	10·8	−0·5	−5·4	0·866	9·35	−0·5	−5·4	−0·866	−9·35
150	11·4	−0·866	−9·87	0·5	5·7	0·5	5·7	−0·866	−9·87
180	9·9	−1	−9·9	0	0	1	9·9	0	0
210	4·8	−0·866	−4·16	−0·5	−2·4	0·5	2·4	0·866	4·16
240	0	−0·5	0	−0·866	0	−0·5	0	0·866	0
270	0	0	0	−1	0	−1	0	0	0
300	0	0·5	0	−0·866	0	−0·5	0	−0·866	0
330	0	0·866	0	−0·5	0	0·5	0	−0·866	0
Σ	53·6		−24·59		27·46		2·10		−8·31
$\Sigma/6$	8·9		−4·1		4·6		0·3		−1·4

$a_0 = 8·9 \qquad a_1 = -4·1 \qquad b_1 = 4·6 \qquad a_2 = 0·3 \qquad b_2 = -1·4$

Now continue the table to find a_3, b_3, a_4 and b_4.

****

50A

$a_3 = -0·8 \qquad b_3 = 0·0 \qquad a_4 = 0·2 \qquad b_4 = 0·1$

FRAME 51

Further calculation gives $a_5 = -0·1, \qquad b_5 = -0·1$

Thus the series is

$4·5 - 4·1 \cos\theta + 0·3 \cos 2\theta - 0·8 \cos 3\theta + 0·2 \cos 4\theta - 0·1 \cos 5\theta ..$
$\qquad + 4·6 \sin\theta - 1·4 \sin 2\theta \qquad\qquad + 0·1 \sin 4\theta - 0·1 \sin 5\theta ....$

This can also be expressed in the form

FRAME 51 continued

$$4 \cdot 5 + 6 \cdot 2 \sin(\theta - 40°) + 1 \cdot 4 \sin(2\theta + 168°) + 0 \cdot 8 \sin(3\theta - 90°)$$
$$+ 0 \cdot 2 \sin(4\theta + 27°) + 0 \cdot 1 \sin(5\theta - 135°) + \ldots$$

You will notice that the amplitudes of successive harmonics decrease so that the higher harmonics contribute very little to the total.

The approximate method we have used for calculating the a's and b's becomes less accurate as the harmonics get higher. The only way of improving the accuracy of the method is to decrease the interval between the tabular values, which means, of course, increasing the number of readings. As a general guide, if the period is divided into k intervals the method can only be relied upon to estimate the first $\left(\frac{k}{2} - 1\right)$ harmonics.

In the present problem, for instance, with 12 intervals we can only expect to go to the fifth harmonic, and even then we would usually only consider the amplitudes of the fourth and fifth harmonics relative to that of the fundamental instead of treating each as being accurate in itself.

In previous frames you have seen what happens to the Fourier series in certain special cases (e.g. odd and even functions) and also how to deal with a period other than 2π. The corresponding results for numerical harmonic analysis will be obvious to you if you remember that it is simply a matter of replacing integration by summation. You will meet some of these cases in the Miscellaneous Examples in FRAME 53.

FRAME 52

Summary

Before giving you some miscellaneous examples to try we will summarise briefly the main results, for easy reference.

NOTE: In each case the formula for a_0 can be obtained by putting n = 0 in that for a_n.

FOURIER SERIES

FRAME 52 continued

Period 2π

$$f(x) = \tfrac{1}{2}a_o + \sum_{n=1}^{\infty} a_n \cos nx + \sum_{n=1}^{\infty} b_n \sin nx$$

Full range: $f(x)$ defined for $-\pi < x < \pi$

$$a_n = \frac{1}{\pi}\int_{-\pi}^{\pi} f(x) \cos nx \, dx$$

$$b_n = \frac{1}{\pi}\int_{-\pi}^{\pi} f(x) \sin nx \, dx$$

Half-range: $f(x)$ defined for $0 < x < \pi$

Cosine series: $a_n = \dfrac{2}{\pi}\int_{0}^{\pi} f(x) \cos nx \, dx$

$b_n = 0$

Sine series: $a_n = 0$

$b_n = \dfrac{2}{\pi}\int_{0}^{\pi} f(x) \sin nx \, dx$

General Period 2ℓ

$$f(x) = \tfrac{1}{2}a_o + \sum_{n=1}^{\infty} a_n \cos \frac{n\pi x}{\ell} + \sum_{n=1}^{\infty} b_n \sin \frac{n\pi x}{\ell}$$

Full range: $f(x)$ defined for $-\ell < x < \ell$

$$a_n = \frac{1}{\ell}\int_{-\ell}^{\ell} f(x) \cos \frac{n\pi x}{\ell} \, dx$$

$$b_n = \frac{1}{\ell}\int_{-\ell}^{\ell} f(x) \sin \frac{n\pi x}{\ell} \, dx$$

Half-range: $f(x)$ defined for $0 < x < \ell$

Cosine series: $a_n = \dfrac{2}{\ell}\int_{0}^{\ell} f(x) \cos \dfrac{n\pi x}{\ell} \, dx$

$b_n = 0$

FRAME 52 continued

Sine Series: $a_n = 0$

$$b_n = \frac{2}{\ell} \int_0^\ell f(x) \sin \frac{n\pi x}{\ell} dx$$

Numerical Harmonic Analysis

For period 2π: a_n = 2 × Mean value of $f(x) \cos nx$ over one period
b_n = 2 ×   "   "   " $f(x) \sin nx$ "   "   "

FRAME 53

Miscellaneous Examples

In this frame a collection of miscellaneous examples is given for you to try. Answers are supplied in FRAME 54, together with such working as is considered helpful. A rough sketch of the waveform is generally advisable.

1. Find the Fourier series for the function defined by:
$$\begin{cases} f(x) = e^x & -\pi < x < \pi \\ f(x + 2\pi) = f(x) \end{cases}$$

NOTE: $\int e^{ax} \sin bx \, dx = \dfrac{e^{ax}(a \sin bx - b \cos bx)}{a^2 + b^2}$

and $\int e^{ax} \cos bx \, dx = \dfrac{e^{ax}(a \cos bx + b \sin bx)}{a^2 + b^2}$

2. Find the Fourier sine series for the trapezoidal wave defined for the half-range 0 to π as shown in the diagram.

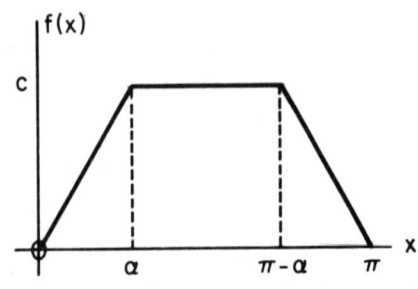

This wave is important in determining the flux distribution in the air gap of an alternator.

FOURIER SERIES

FRAME 53 continued

3. On a long beam, patches of constant loading p_0 of length $k\ell$ alternate with patches of length $(1 - k)\ell$ free of pressure, as shown in the diagram.

Find the Fourier analysis for the loading. By putting $k = \frac{1}{2}$ in this series, verify the result (7.1).

4. Find a Fourier series to represent the rectified sine wave $e = E|\sin \omega t|$. (This is the output voltage when the e.m.f. $E \sin \omega t$ acts on a full-wave rectifier.)

5. Find (i) a half-range Fourier sine series, and
        (ii) a half-range Fourier cosine series,
   to represent the function defined between 0 and T as $y = 1 + t/T$.

6. A function $f(x)$ of period 2ℓ is defined by
   $$f(x) = -\frac{k}{\ell}(\ell + x) \qquad -\ell < x < 0$$
   $$f(x) = \frac{k}{\ell}(\ell - x) \qquad 0 < x < \ell$$
   Find the Fourier series which represents this.

The following three examples are all numerical. As the technique involved is similar in each case, it is suggested that you try only one of them, choosing one which is nearest to your own particular interest.

7. Referring to FRAME 38, Figure (iv), the voltage produced by one such machine is as shown in the table, the period being divided into twelve equal intervals.

FOURIER SERIES

FRAME 53 continued

θ°	0	30	60	90	120	150	180	210	240	270	300	330
v	137	164	265	325	156	−54	−137	−164	−265	−325	−156	54

Analyse this waveform as far as the fifth harmonic.

8. A person's temperature is a function of the time of day and his temperature curve normally repeats itself daily. Thus his temperature is the same at, for example, 2 pm every day. The following values were extracted from a one-hourly temperature chart.

Time	Mid-night	2 am	4 am	6 am	8 am	10 am	Noon	2 pm	4 pm	6 pm	8 pm	10 pm
Temp °C	36·61	36·50	36·45	36·54	36·69	36·76	36·94	37·13	37·35	37·25	36·98	36·86

As in the example in FRAME 50 the table shows 12 readings taken at equal intervals over one cycle (midnight corresponds to $\theta = 0$, 2 am to $\theta = \pi/6$ etc).

Find a Fourier series as far as the third harmonic to represent the temperature curve.

9. The table shows the displacement x mm of a sliding piece from a fixed reference point for every 30° of rotation of the crank.

θ	0	30	60	90	120	150	180	210	240	270	300	330
x	298	356	373	337	254	155	80	51	60	93	147	221

Find a Fourier series for x as far as the third harmonic.

FOURIER SERIES

FRAME 54

Answers to Miscellaneous Examples

1.

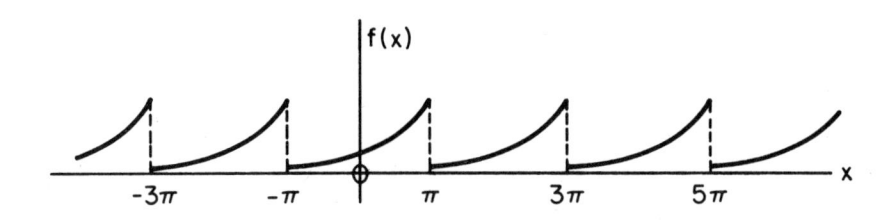

[Graph of $f(x)$ showing a periodic sawtooth-like function with exponential rise on intervals $(-3\pi, -\pi)$, $(-\pi, \pi)$, $(\pi, 3\pi)$, $(3\pi, 5\pi)$, with marks at $-3\pi, -\pi, \pi, 3\pi, 5\pi$ on the x-axis.]

$$a_o = \frac{1}{\pi}\int_{-\pi}^{\pi} e^x\, dx = \frac{1}{\pi}(e^\pi - e^{-\pi}) = \frac{2\sinh \pi}{\pi}$$

$$a_n = \frac{1}{\pi}\int_{-\pi}^{\pi} e^x \cos nx\, dx$$

$$= \frac{1}{\pi}\left[\frac{e^x}{1+n^2}(\cos nx + n\sin nx)\right]_{-\pi}^{\pi}$$

$$= \frac{1}{\pi(1+n^2)}\left[e^\pi \cos n\pi - e^{-\pi}\cos n\pi\right]$$

$$= \frac{e^\pi - e^{-\pi}}{\pi(1+n^2)}\cos n\pi$$

$$= \begin{cases} \dfrac{2\sinh \pi}{\pi(1+n^2)} & \text{if } n \text{ is even} \\[6pt] \dfrac{-2\sinh \pi}{\pi(1+n^2)} & \text{if } n \text{ is odd} \end{cases}$$

$$b_n = \frac{1}{\pi}\int_{-\pi}^{\pi} e^x \sin nx\, dx$$

$$= \begin{cases} \dfrac{-2n\sinh \pi}{\pi(1+n^2)} & \text{if } n \text{ is even} \\[6pt] \dfrac{2n\sinh \pi}{\pi(1+n^2)} & \text{if } n \text{ is odd} \end{cases}$$

$$\therefore f(x) = \frac{2\sinh \pi}{\pi}\left\{\frac{1}{2} - \frac{1}{2}\cos x + \frac{1}{5}\cos 2x - \frac{1}{10}\cos 3x + \cdots \right. $$
$$\left. + \frac{1}{2}\sin x - \frac{2}{5}\sin 2x + \frac{3}{10}\sin 3x - \cdots\right\}$$

2.

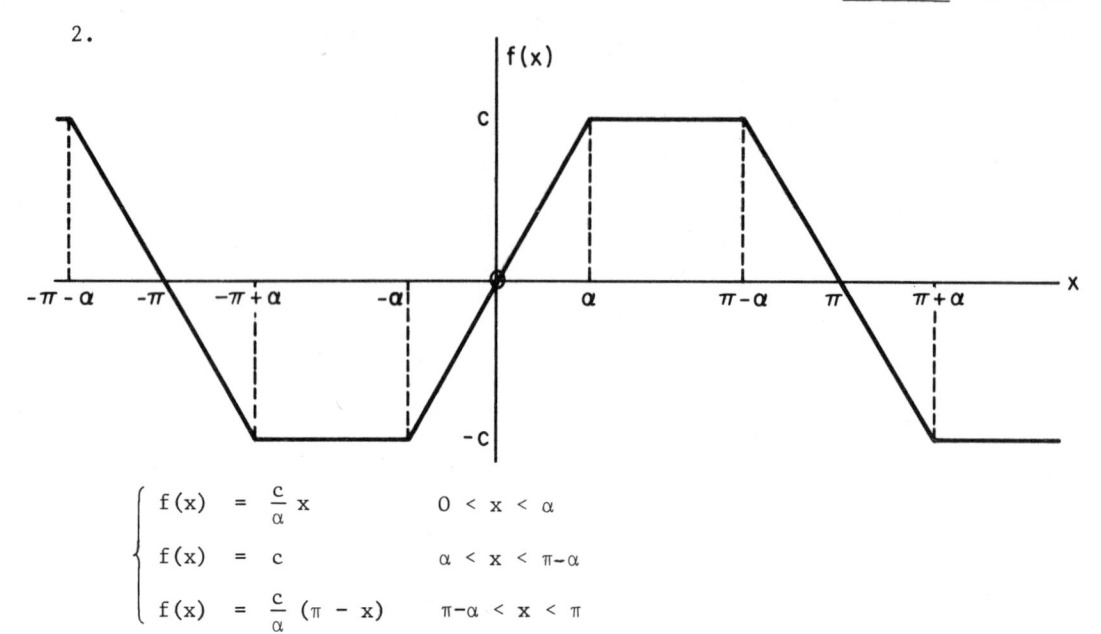

$$\begin{cases} f(x) = \dfrac{c}{\alpha} x & 0 < x < \alpha \\ f(x) = c & \alpha < x < \pi-\alpha \\ f(x) = \dfrac{c}{\alpha}(\pi - x) & \pi-\alpha < x < \pi \end{cases}$$

There will only be odd harmonics.

$$b_n = \frac{2}{\pi}\left[\int_0^\alpha \frac{c}{\alpha} x \sin nx\, dx + \int_\alpha^{\pi-\alpha} c \sin nx\, dx + \int_{\pi-\alpha}^\pi \frac{c}{\alpha}(\pi - x) \sin nx\, dx\right]$$

$$= \frac{2c}{\pi n^2 \alpha}\left[\sin n\alpha + \sin n(\pi - \alpha)\right]$$

$$= \begin{cases} 0 & \text{if } n \text{ is even} \\ \dfrac{4c \sin n\alpha}{\pi n^2 \alpha} & \text{if } n \text{ is odd} \end{cases}$$

$$f(x) = \frac{4c}{\pi\alpha}\left(\sin \alpha \sin x + \frac{\sin 3\alpha}{3^2}\sin 3x + \frac{\sin 5\alpha}{5^2}\sin 5x + \ldots\right)$$

3.
$$\begin{cases} f(x) = p_o & 0 < x < k\ell \\ f(x) = 0 & k\ell < x < \ell \\ \text{Period } \ell \end{cases}$$

$$a_o = \frac{2}{\ell} \times p_o\, k\ell = 2kp_o$$

FOURIER SERIES

FRAME 54 continued

$$a_n = \frac{2}{\ell} \int_0^{k\ell} p_o \cos \frac{2n\pi x}{\ell} dx = \frac{p_o}{n\pi} \sin 2nk\pi$$

$$b_n = \frac{2}{\ell} \int_0^{k\ell} p_o \sin \frac{2n\pi x}{\ell} dx = \frac{p_o}{n\pi} (1 - \cos 2nk\pi)$$

$$f(x) = p_o k + \frac{p_o}{\pi}\left[\sin 2k\pi \cos \frac{2\pi x}{\ell} + \frac{1}{2}\sin 4k\pi \cos \frac{4\pi x}{\ell} + \frac{1}{3}\sin 6k\pi \cos \frac{6\pi x}{\ell} + ..\right.$$

$$\left. + (1-\cos 2k\pi) \sin \frac{2\pi x}{\ell} + \frac{1}{2}(1-\cos 4k\pi) \sin \frac{4\pi x}{\ell} + \frac{1}{3}(1-\cos 6k\pi) \sin \frac{6\pi x}{\ell} + ..\right]$$

Putting $k = \frac{1}{2}$ gives the series

$$\frac{1}{2} p_o + \frac{2p_o}{\pi}\left[\sin \frac{2\pi x}{\ell} + \frac{1}{3} \sin \frac{6\pi x}{\ell} + \frac{1}{5} \sin \frac{10\pi x}{\ell} + ...\right]$$

which was quoted for the patch loading in FRAME 7.

4.

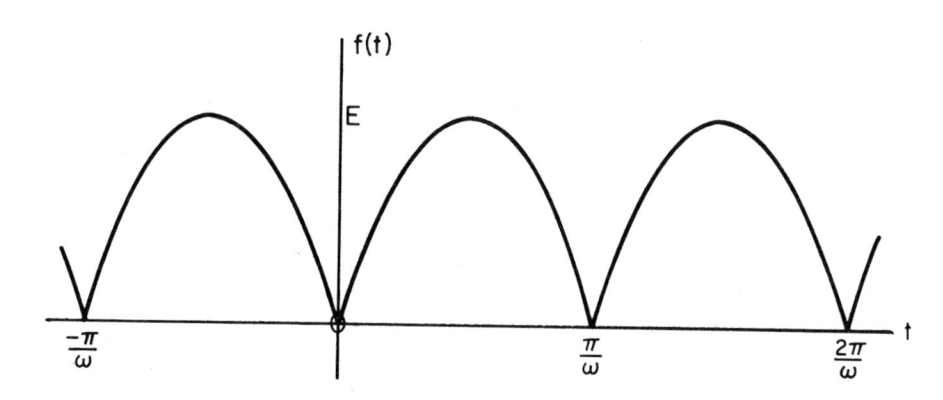

This is an even function, period π/ω.

$$a_o = \frac{2\omega}{\pi} \int_0^{\pi/\omega} E \sin \omega t \, dt$$

$$= 4E/\pi$$

FRAME 54 continued

$$a_n = \frac{2\omega}{\pi} \int_0^{\pi/\omega} E \sin \omega t \cos 2n\omega t \, dt$$

$$= \frac{E\omega}{\pi} \int_0^{\pi/\omega} \{\sin(2n+1)\omega t - \sin(2n-1)\omega t\} dt$$

$$= -\frac{4E}{\pi} \frac{1}{(2n-1)(2n+1)}$$

Did you integrate by parts? If so you were using an unnecessarily complicated method.

$$e = \frac{2E}{\pi} - \frac{4E}{\pi} \left(\frac{1}{1.3} \cos 2\omega t + \frac{1}{3.5} \cos 4\omega t + \frac{1}{5.7} \cos 6\omega t + \ldots \right)$$

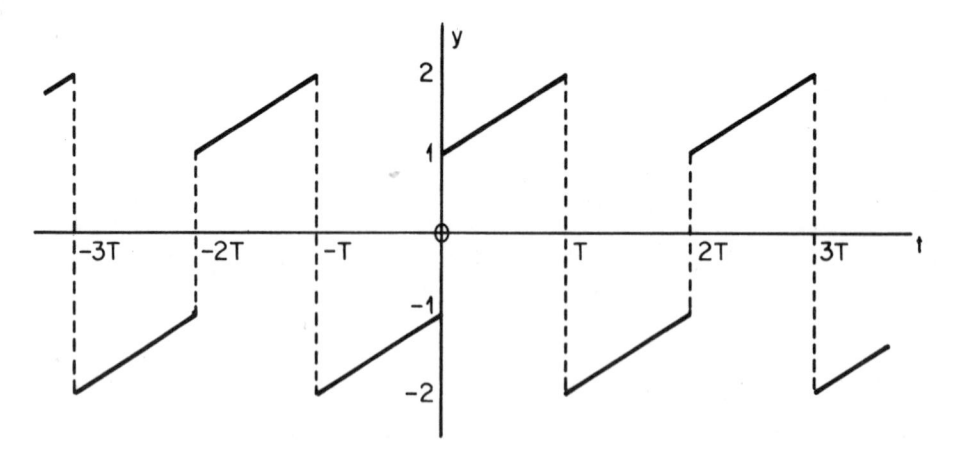

5. (i)

$$b_n = \frac{2}{T} \int_0^T (1 + \frac{t}{T}) \sin \frac{n\pi t}{T} dt$$

$$= \frac{2}{n\pi} (1 - 2 \cos n\pi)$$

$$= \begin{cases} \frac{-2}{n\pi} & \text{if } n \text{ is even} \\ \frac{6}{n\pi} & \text{if } n \text{ is odd} \end{cases}$$

$$y = \frac{2}{\pi} \left(3 \sin \frac{\pi t}{T} - \frac{1}{2} \sin \frac{2\pi t}{T} + \frac{3}{3} \sin \frac{3\pi t}{T} - \frac{1}{4} \sin \frac{4\pi t}{T} + \frac{3}{5} \sin \frac{5\pi t}{T} - \ldots \right)$$

FOURIER SERIES

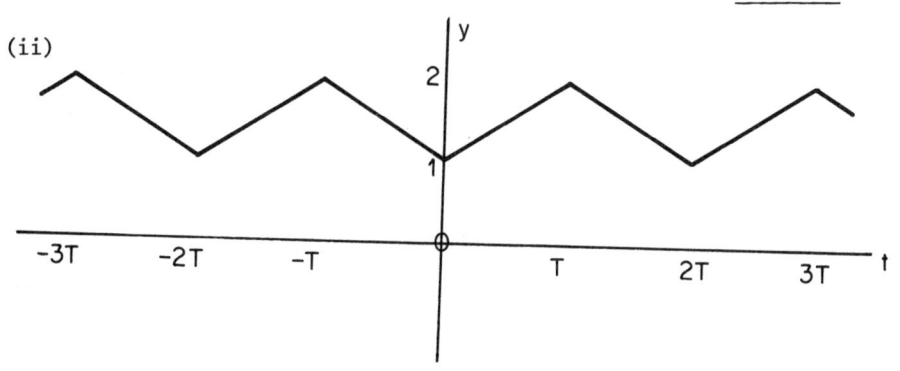

FRAME 54 continued

(ii)

$$a_o = \frac{2}{T} \int_o^T (1 + \frac{t}{T}) dt$$

$$= 3$$

$$a_n = \frac{2}{T} \int_o^T (1 + \frac{t}{T}) \cos \frac{n\pi t}{T} dt$$

$$= \frac{2}{\pi^2 n^2} (\cos n\pi - 1)$$

$$= \begin{cases} 0 & \text{if } n \text{ is even} \\ \frac{-4}{\pi^2 n^2} & \text{if } n \text{ is odd} \end{cases}$$

$$y = \frac{3}{2} - \frac{4}{\pi^2} \left[\cos \frac{\pi t}{T} + \frac{1}{3^2} \cos \frac{3\pi t}{T} + \frac{1}{5^2} \cos \frac{5\pi t}{T} + \ldots \right]$$

6.

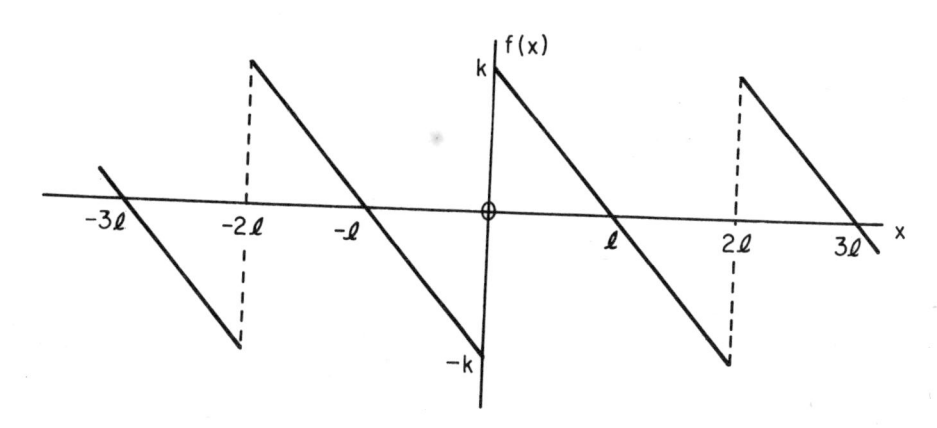

FRAME 54 continued

This is an odd function. It can be expanded either as a full-range series, working from 0 to 2ℓ, or as a half-range series. Using the latter method:

$$b_n = \frac{2}{\ell} \int_0^\ell \frac{k}{\ell} (\ell - x) \sin \frac{n\pi x}{\ell} \, dx$$

$$= \frac{2k}{n\pi}$$

$$f(x) = \frac{2k}{\pi} \left(\sin \frac{\pi x}{\ell} + \frac{1}{2} \sin \frac{2\pi x}{\ell} + \frac{1}{3} \sin \frac{3\pi x}{\ell} + \ldots \right)$$

7. As $f(\theta + 180°) = -f(\theta)$, only the odd harmonics are present so it is only necessary to calculate a_1, a_3, a_5, b_1, b_3 and b_5.

$$\begin{aligned} v &= 127 \cos\theta + 9 \cos 3\theta + \cos 5\theta \\ &\quad + 248 \sin\theta - 72 \sin 3\theta + 5 \sin 5\theta \\ &= 279 \sin(\theta + 27°) + 72 \sin(3\theta + 173°) + 5 \sin(5\theta + 11°) \end{aligned}$$

Did you notice that the values for $v \cos\theta$, $v \sin\theta$, $v \cos 3\theta$ etc. for $\theta = 180$ to 330 repeat those for $\theta = 0$ to 150? This always happens if only the odd harmonics are present and consequently only the table from 0 to 150 is really needed.

8. $36 \cdot 84 - 0 \cdot 18 \cos\theta - 0 \cdot 06 \cos 2\theta + 0 \cdot 05 \cos 3\theta$
    $\quad\quad\quad - 0 \cdot 35 \sin\theta + 0 \cdot 02 \sin 2\theta$

This could be given in terms of t (hours measured from midnight) and would then be

$36 \cdot 84 - 0 \cdot 18 \cos \frac{\pi t}{12} - 0 \cdot 06 \cos \frac{2\pi t}{12} + 0 \cdot 05 \cos \frac{3\pi t}{12}$
$\quad\quad\quad - 0 \cdot 35 \sin \frac{\pi t}{12} + 0 \cdot 02 \sin \frac{2\pi t}{12}$

However, it is the coefficients which are the main interest, as different medical cases show different patterns for these.

9. $x = 202 + 107 \cos\theta - 13 \cos 2\theta + 2 \cos 3\theta$
    $\quad\quad\quad + 121 \sin\theta + 9 \sin 2\theta - \sin 3\theta$

PARTIAL DIFFERENTIAL EQUATIONS FOR TECHNOLOGISTS

A PROGRAMMED TEXT

I. M. Calus
J. A. Fairley

PREFACE

Although there are many different types of partial differential equations with a variety of methods of solution, it is mainly one particular kind which occurs in engineering and applied science. This programme will concentrate on the solution of such equations by one method (separation of variables), this method being of the most value in practice.

INSTRUCTIONS

This programme constitutes a self-instructional course on the solution of certain partial differential equations mainly by the method of separation of variables

The programme is divided up into a number of FRAMES which are to be worked *in the order given*. You will be required to participate in many of these frames and in such cases hints and answers are provided in ANSWER FRAMES, designated by the letter A following the frame number. The answer frame is separated from the main frame by a line of asterisks: ******************. Keep the answers covered until you have written your own response. Do not proceed to the next frame until you have corrected any mistakes in your attempt and are satisfied that you understand the contents up to this point.

Before reading this programme, it is necessary that you are familiar with the following

Prerequisites

Partial differential coefficients and the total differential. Ordinary differential equations — linear with constant coefficients and R.H.S. zero. (These are dealt with in Programme 2 of the book *Ordinary Differential Equations* by A. C. Bajpai, I. M. Calus and J. Hyslop.)

Fourier series (see Programme 1 of this volume).

PARTIAL DIFFERENTIAL EQUATIONS

FRAME 1

Introduction

If y is a function of one variable x, an equation involving its derivatives will be an ordinary differential equation,

e.g. $\quad \dfrac{d^2y}{dx^2} + 6\dfrac{dy}{dx} - 2y = \sin 3x \qquad (1.1)$

If, however, y is a function of two variables x and t, any derivatives with respect to x or t will be partial and so an equation involving them will be a partial differential equation,

e.g. $\quad \dfrac{\partial^2 y}{\partial t^2} + 6\dfrac{\partial y}{\partial t} = 4\dfrac{\partial^2 y}{\partial x^2} \qquad (1.2)$

The same is true for a function of more than two variables, so that we might have

$$\dfrac{\partial^2 \theta}{\partial x^2} + \dfrac{\partial^2 \theta}{\partial y^2} + \dfrac{\partial^2 \theta}{\partial z^2} = \dfrac{1}{h^2}\dfrac{\partial \theta}{\partial t} \qquad (1.3)$$

where θ is a function of x, y, z and t (h is a constant).

The solution of (1.1) involves finding y as a function of x, that of (1.2) y as a function of x and t and that of (1.3) θ as a function of x, y, z and t.

This programme will be restricted to the solution of partial differential equations involving two independent variables only.

An example of such an equation is (1.2). Like all differential equations it has an unlimited number of solutions. One of these is
$$y = Ae^{-3t}\sin\dfrac{3x}{2}.$$
You might like to verify this.

****

1A

$\dfrac{\partial y}{\partial t} = -3Ae^{-3t}\sin\dfrac{3x}{2}$

$\dfrac{\partial^2 y}{\partial t^2} = 9Ae^{-3t}\sin\dfrac{3x}{2}$

$\dfrac{\partial^2 y}{\partial x^2} = -\dfrac{9}{4}Ae^{-3t}\sin\dfrac{3x}{2}$

$\dfrac{\partial^2 y}{\partial t^2} + 6\dfrac{\partial y}{\partial t} = -9Ae^{-3t}\sin\dfrac{3x}{2} = 4\dfrac{\partial^2 y}{\partial x^2}$

FRAME 2

We shall now mention a few cases where partial differential equations arise in practice.

One example is that of the transverse vibration of a tightly stretched string.

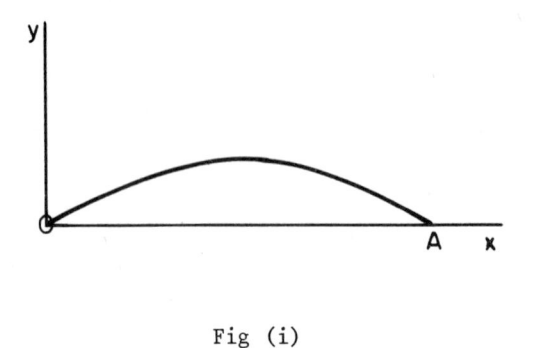

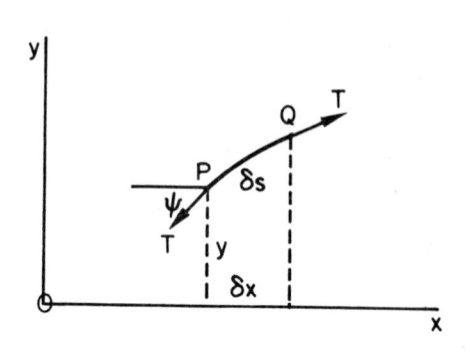

Fig (i)

Fig (ii)

Suppose that the ends of the string are fixed at the two points O and A and that the string is set in motion in the plane Oxy. At a subsequent time t let the string be displaced as shown in Fig (i). It is assumed that the amplitude of the oscillations is small. Therefore the change in length of the string during the motion is small enough for the tension T to be assumed constant. Fig (ii) shows a small element PQ of the string on an enlarged scale.

The acceleration of this element is $\frac{\partial^2 y}{\partial t^2}$, the partial derivative being used as y is a function of both x and t. Hence the force required to produce this acceleration is $\rho \delta s \frac{\partial^2 y}{\partial t^2}$ where ρ is the mass per unit length of the string.

The actual force producing the motion is the difference between the components parallel to Oy of T at the points P and Q, i.e. the increase in the component of T in this direction as x increases by an amount δx.

At P, this component of T is $T \sin \psi$.

Now $\sin \psi = \psi - \frac{\psi^3}{3!} \ldots$ and $\tan \psi = \psi + \frac{\psi^3}{3} \ldots$, and as ψ is small, powers of ψ can be neglected.

FRAME 2 continued

Hence $\sin \psi \simeq \tan \psi = \frac{\partial y}{\partial x}$ and the component of T at P is $T \frac{\partial y}{\partial x}$.

The increase in this as x increases by δx is $\frac{\partial}{\partial x}\left(T \frac{\partial y}{\partial x}\right) \delta x$

$$= T \frac{\partial^2 y}{\partial x^2} \delta x$$

∴ The equation of motion is $\rho \delta s \frac{\partial^2 y}{\partial t^2} = T \frac{\partial^2 y}{\partial x^2} \delta x$

Now $\delta x = \delta s \cos \psi$

$\simeq \delta s$ as $\cos \psi \simeq 1$ if powers of ψ are ignored.

∴ $\rho \frac{\partial^2 y}{\partial t^2} = T \frac{\partial^2 y}{\partial x^2}$

This is usually written $\frac{\partial^2 y}{\partial t^2} = c^2 \frac{\partial^2 y}{\partial x^2}$ where $c^2 = T/\rho$ and is constant.

FRAME 3

Another example occurs in the study of simple fluid flow.

Suppose a non-viscous fluid of unit depth is streaming over a horizontal plane and axes Ox, Oy are taken anywhere in the plane. Let the velocity of the fluid at any point whose horizontal coordinates are x and y have components u and v parallel to Ox and Oy.

Consider an imaginary rectangle with sides δx and δy, and centre (x,y) as shown in the diagram.

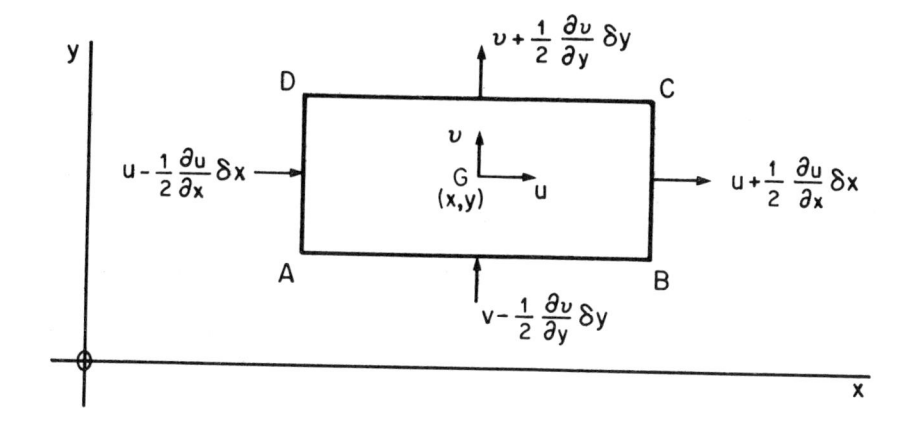

FRAME 3 continued

The rate of increase of u with respect to x is $\frac{\partial u}{\partial x}$.

∴ The increase in u in going from G to E is $\frac{\partial u}{\partial x} \cdot \frac{1}{2} \delta x$ and hence the velocity component at E parallel to Ox is $u + \frac{1}{2} \frac{\partial u}{\partial x} \delta x$.

Similarly the velocity components at the mid-points of the other sides are as shown in the diagram.

Thus in unit time the volume of fluid crossing over BC is approximately $\left(u + \frac{1}{2} \frac{\partial u}{\partial x} \delta x\right) \delta y$ and over AD is $\left(u - \frac{1}{2} \frac{\partial u}{\partial x} \delta x\right) \delta y$, giving between them a net loss of $\left(u + \frac{1}{2} \frac{\partial u}{\partial x} \delta x\right) \delta y - \left(u - \frac{1}{2} \frac{\partial u}{\partial x} \delta x\right) \delta y$

$$= \frac{\partial u}{\partial x} \delta x \delta y$$

Similarly, for the flow over AB and DC, the net loss is $\frac{\partial v}{\partial y} \delta y \delta x$.

∴ Total net loss = $\left(\frac{\partial u}{\partial x} + \frac{\partial v}{\partial y}\right) \delta x \delta y$.

If there is no sink or source within the rectangle, and the fluid is assumed to be of constant density, this must be zero.

$$\therefore \frac{\partial u}{\partial x} + \frac{\partial v}{\partial y} = 0 \qquad (3.1)$$

If the flow is irrotational, it can be shown that

$$u = \frac{\partial \phi}{\partial x} \quad \text{and} \quad v = \frac{\partial \phi}{\partial y}$$

where ϕ is a function called the velocity potential.

Equation (3.1) then becomes

$$\frac{\partial^2 \phi}{\partial x^2} + \frac{\partial^2 \phi}{\partial y^2} = 0$$

This is Laplace's Equation in two dimensions and also occurs in other cases of continuous movement such as heat flow and magnetic or electric flux.

FOR TECHNOLOGISTS

FRAME 4

Some more examples are listed below. However, in these cases, only the final differential equation will be quoted, the derivation being omitted.

(i) If the sides of a bar of constant cross-sectional area are insulated and heat flows along the bar then the temperature θ at a point distant x from one end at time t satisfies the equation

$$\frac{\partial^2 \theta}{\partial x^2} = \frac{1}{h^2} \frac{\partial \theta}{\partial t}$$

where h is a constant.

(ii) If a uniform beam vibrates transversely then the displacement y at a distance x from one end at time t satisfies the equation

$$EI \frac{\partial^4 y}{\partial x^4} + m \frac{\partial^2 y}{\partial t^2} = 0$$

(iii) If a transmission line consists of two parallel wires, the voltage drop v across the wires and the current i in the line, at a distance x from the sending end at time t, satisfy the equations

$$CL \frac{\partial^2 v}{\partial t^2} + (CR + LG) \frac{\partial v}{\partial t} + RGv = \frac{\partial^2 v}{\partial x^2}$$

$$\text{and} \quad CL \frac{\partial^2 i}{\partial t^2} + (CR + LG) \frac{\partial i}{\partial t} + RGi = \frac{\partial^2 i}{\partial x^2}$$

where C, L, R and G are respectively the capacitance, inductance, resistance and leakance per unit length.

FRAME 5

Solution by direct integration

You will remember that the simplest ordinary differential equations such as $\frac{d^2 y}{dx^2} = f(x)$ can be solved by direct integration.

For instance, if $\frac{d^2 y}{dx^2} = 2$

FRAME 5 continued

$$\text{then} \quad \frac{dy}{dx} = 2x + A$$

$$\text{and} \quad y = x^2 + Ax + B$$

where A and B are constants, each of which can take any value. Here, $2x + A$ is the most general function that gives 2 when differentiated with respect to x and $x^2 + Ax + B$ is the most general function that gives $2x + A$. $y = x^2 + Ax + B$ is the general solution of the differential equation $\frac{d^2y}{dx^2} = 2$.

In a similar manner certain partial differential equations can be solved by direct integration,

$$\text{e.g.} \quad \frac{\partial^2 y}{\partial x^2} = 2 \quad \text{where y is a function of x and t.}$$

Then, $\frac{\partial y}{\partial x} = 2x + f(t)$ because $2x + f(t)$ is the most general function that gives 2 when differentiated partially with respect to x.

In the same way, $y = x^2 + x f(t) + g(t)$ and this is the general solution, $f(t)$ and $g(t)$ being arbitrary functions of t (which may contain constants).

Now find the general solution of

$$\frac{\partial^2 y}{\partial x \partial t} = 2$$

****

5A

$$\frac{\partial y}{\partial t} = 2x + f(t)$$

$$y = 2xt + \int f(t)\,dt + g(x)$$

$$= 2xt + F(t) + g(x)$$

$f(t)$ being an arbitrary function of t, $\int f(t)\,dt$ is also an arbitrary function, which we have called $F(t)$.

FRAME 6

Returning to the equation $\frac{d^2y}{dx^2} = 2$ in FRAME 5, if additional information is given, e.g. when $x = 0$, $y = -3$ and $\frac{dy}{dx} = 1$, it is possible to determine the values of A and B.

In the case of the partial differential equation $\frac{\partial^2 y}{\partial x^2} = 2$ (see FRAME 5 again) it is possible to find suitable expressions for $f(t)$ and $g(t)$ if additional information is available.

For instance, if $y = \sin t$ for all t when $x = 0$ and $y = x^2 + 2x$ for all x when $t = 0$, substitution in the general solution
$$y = x^2 + x f(t) + g(t)$$
gives
$$\sin t = g(t)$$
and $x^2 + 2x = x^2 + x f(0) + g(0)$
$\therefore \quad 2x = x f(0) \quad$ as $\quad g(0) = 0$
$\therefore \quad f(0) = 2$

Thus $f(t)$ is any function of t which has the value 2 when $t = 0$. Write down a few expressions for $f(t)$ which satisfy this condition.

****

6A

Some possible expressions are:
$$2, \quad 2 + t, \quad 2 \cos t, \quad 2e^t, \quad 2 \cosh t.$$
There are, of course, many more.

FRAME 7

As any of these expressions for $f(t)$ is suitable, one solution is
$$y = x^2 + 2x + \sin t$$
Another is $y = x^2 + 2x \cos t + \sin t$, and so on.
In FRAME 5 you found the general solution of $\frac{\partial^2 y}{\partial x \partial t} = 2$. Now see if you can find suitable expressions for $F(t)$ and $g(x)$ so that $y = 0$ when $x = 0$ for all t, and $y = x^2$ when $t = 0$ for all x.

****

7A

$$y = 0 \text{ when } x = 0 \quad \therefore \; 0 = F(t) + g(0) \quad (7A.1)$$

$$y = x^2 \text{ when } t = 0 \quad \therefore \; x^2 = F(0) + g(x) \quad (7A.2)$$

Equation (7A.2) is satisfied by $g(x) = x^2 - A$, $F(0) = A$ where A is a constant.

Then $g(0) = -A$ and (7A.1) gives $F(t) = A$

$$\therefore \quad y = 2xt + A + x^2 - A$$
$$= 2xt + x^2$$

You will notice that, in this case, although there are many possibilities for $F(t)$ and $g(x)$, there is only one solution for y.

FRAME 8

Two simple cases where partial differential equations are solved by direct integration are:

(i)   The solution of a certain type of ordinary differential equation i.e. first order exact (see Programme 1 in "Ordinary Differential Equations" by A.C. Bajpai, I.M. Calus and J. Hyslop).

(ii) Conjugate functions.

Taking first an example of (i), the equation

$$(2xy - \sin y)dx + (x^2 + 3y^2 - x \cos y)dy = 0 \quad (8.1)$$

is exact, i.e. it is of the form $\frac{\partial u}{\partial x} dx + \frac{\partial u}{\partial y} dy = 0$. This is equivalent to $du = 0$, the solution of which is $u = c$.

Here, $\frac{\partial u}{\partial x} = 2xy - \sin y$ and $\frac{\partial u}{\partial y} = x^2 + 3y^2 - x \cos y$

Integrating, these two equations give respectively

$$u = x^2 y - x \sin y + f(y)$$
$$\text{and} \quad u = x^2 y + y^3 - x \sin y + g(x)$$

FOR TECHNOLOGISTS

FRAME 8 continued

Comparing these, you will see that these two requirements are met by $f(y) = y^3$, $g(x) = 0$. The solution of (8.1) is

$$x^2 y - x \sin y + y^3 = c$$

(The technique used here is slightly different from that used in the programme mentioned above.)

Now solve the exact equation

$$(2x - e^y)dx + (2 - xe^y)dy = 0$$

****

8A

$\dfrac{\partial u}{\partial x} = 2x - e^y$ and $\dfrac{\partial u}{\partial y} = 2 - xe^y$

$u = x^2 - xe^y + f(y)$ and

$u = 2y - xe^y + g(x)$

Comparing, $f(y) = 2y$ and $g(x) = x^2$

The solution is $x^2 - xe^y + 2y = c$

FRAME 9

Two functions u and v, of x and y, are called conjugate if they satisfy the equations

$$\frac{\partial u}{\partial x} = \frac{\partial v}{\partial y} \quad \text{and} \quad \frac{\partial u}{\partial y} = -\frac{\partial v}{\partial x}$$

Such functions occur in the study of heat flow, electrostatics and fluid motion. In heat flow, the equations $u = \alpha$, $v = \beta$, represent the isothermal lines and flux lines respectively; in electrostatics they represent equipotential lines and flux lines, and in fluid motion equipotential lines and stream lines. Given one of these functions, it is possible to find the other. If $v = 2xy$, see if you can find u.

****

$$\frac{\partial u}{\partial x} = 2x \quad and \quad \frac{\partial u}{\partial y} = -2y$$

$$u = x^2 + f(y)$$

$$u = -y^2 + g(x)$$

$$\therefore u = x^2 - y^2$$

An arbitrary constant can be added to the right hand side.

FRAME 10

An exponential trial solution

You will remember that an exponential trial solution is often useful when dealing with ordinary differential equations.

For example, a trial solution of the form $y = Ae^{mx}$ is the basis of the auxiliary equation when solving

$$a\frac{d^2y}{dx^2} + b\frac{dy}{dx} + cy = 0 \qquad (10.1)$$

(A full treatment of this is given in Programme 2 of "Ordinary Differential Equations" by A.C. Bajpai, I.M. Calus and J. Hyslop.)

A similar technique can sometimes be used in solving partial differential equations.

We shall illustrate this by finding a solution for equation (1.2)

$$i.e. \quad \frac{\partial^2 y}{\partial t^2} + 6\frac{\partial y}{\partial t} = 4\frac{\partial^2 y}{\partial x^2}$$

which is analogous to (10.1) in that it is linear with constant coefficients.

Taking as a trial solution $y = Ae^{mx+nt}$

$$\frac{\partial y}{\partial t} = nAe^{mx+nt} \qquad \frac{\partial^2 y}{\partial t^2} = n^2Ae^{mx+nt}$$

$$\frac{\partial y}{\partial x} = mAe^{mx+nt} \qquad \frac{\partial^2 y}{\partial x^2} = m^2Ae^{mx+nt}$$

FOR TECHNOLOGISTS   2:11

                                                                    FRAME 10 continued

∴ This is a solution if $n^2 A e^{mx+nt} + 6nAe^{mx+nt} = 4m^2 A e^{mx+nt}$

   i.e. if $n^2 + 6n = 4m^2$

   i.e. if $m = \pm \tfrac{1}{2}\sqrt{n(n+6)}$

A few possible solutions, obtained by giving different values to n, are

$$y = A e^{2x+2t}$$
$$y = A e^{-2x-8t}$$
$$y = A e^{(3/2)ix-3t}$$
$$y = A e^{-(3/2)ix-3t}$$

Now say which of the following are also solutions:

(i) $y = A e^{2x-8t}$

(ii) $y = A e^{4x}$

(iii) $y = A e^{-6t}$

(iv) $y = A e^{-2x+2t}$

****

                                                                                 10A

(i), (iii) and (iv) are solutions.
(ii) is not a solution.

                                                                         FRAME 11

As the equation (1.2) is linear with constant coefficients, the sum of any two
or more solutions is also a solution. For example, each of the following is
also a solution.

$$y = A_1 e^{2x+2t} + A_2 e^{2x-8t}$$
$$y = A_1 e^{-2x+2t} + A_2 e^{-(3/2)ix-3t} + A_3 e^{-6t}$$
$$y = A_1 e^{(3/2)ix-3t} + A_2 e^{-(3/2)ix-3t}$$

FRAME 11 continued

You will realise that A_1, A_2, A_3 have been used instead of A because the coefficients need not all be the same.

The last solution can also be written as

$$y = e^{-3t}\left(A_1 e^{3ix/2} + A_2 e^{-3ix/2}\right)$$

$$= e^{-3t}\left(B_1 \cos \frac{3x}{2} + B_2 \sin \frac{3x}{2}\right)$$

If B_1, which is an arbitrary constant, is put equal to zero, the solution given in FRAME 1 is obtained.

Find the relation between m and n for $y = Ae^{mx+nt}$ to be a solution of $\frac{\partial^2 y}{\partial x^2} + 2\frac{\partial^2 y}{\partial x \partial t} = y$. Use this relation to say which of the following are solutions:

(i) $Ae^{2x - \frac{3}{4}t}$

(ii) Ae^{-x}

(iii) Ae^{-2x+t}

(iv) $A_1 e^x + A_2 e^{i(x-t)}$

(v) $A_1 e^{-\frac{1}{2}x - \frac{3}{4}t} + A_2 e^{3x+it} + A_3 e^{-x}$

****

11A

$m^2 + 2mn = 1$

(i), (ii) and (iv) are solutions.

(iii) and (v) are not solutions.

FOR TECHNOLOGISTS

FRAME 12

The method of separation of variables

The trial solution $y = Ae^{mx+nt}$ can also be written as $y = Ae^{mx}e^{nt}$ which is the product of a function of x and a function of t, i.e. it is of the form $X(x)T(t)$. It is often simpler to start with this as a trial solution rather than the more specific exponential function.

As a first example, we shall find a solution of

$$\frac{\partial^2 y}{\partial x^2} + \frac{\partial^2 y}{\partial t^2} = 0 \qquad (12.1)$$

which satisfies the conditions $y = \sin t$ when $x = 0$ for all t, and $y \to 0$ as $x \to \infty$.

Let a trial solution be $y = X(x)T(t)$, which for convenience we shall write as $y = XT$, where it is understood that X is a function of x only and T is a function of t only.

Then $\frac{\partial y}{\partial x} = T \frac{dX}{dx}$ and $\frac{\partial^2 y}{\partial x^2} = T \frac{d^2 X}{dx^2}$,

$\frac{\partial y}{\partial t} = X \frac{\partial T}{\partial t}$ and $\frac{\partial^2 y}{\partial t^2} = X \frac{d^2 T}{dt^2}$.

Thus $y = XT$ is a solution of (12.1) if

$$T \frac{d^2 X}{dx^2} + X \frac{d^2 T}{dt^2} = 0$$

Separating the variables, this can be written

$$\frac{1}{X} \frac{d^2 X}{dx^2} = -\frac{1}{T} \frac{d^2 T}{dt^2} \qquad (12.2)$$

What would the left-hand side become in each of the following cases?

(i) $X = 4e^{2x}$

(ii) $X = x^2$

(iii) $X = xe^{-x}$

(iv) $X = A \cos 3x + B \sin 3x$

****

2:14   PARTIAL DIFFERENTIAL EQUATIONS

12A

(i)   4   (ii) $\dfrac{2}{x^2}$

(iii) $\dfrac{x-2}{x}$ (iv) -9

FRAME 13

Whilst it would be possible to find T such that $-\dfrac{1}{T}\dfrac{d^2T}{dt^2} = 4$ or -9, it is impossible for $-\dfrac{1}{T}\dfrac{d^2T}{dt^2}$ to be equal to $\dfrac{2}{x^2}$ or $\dfrac{x-2}{x}$, as T contains t only. Consequently you will see that (12.2) can only be satisfied if $\dfrac{1}{X}\dfrac{d^2X}{dx^2}$ is a constant (and therefore $-\dfrac{1}{T}\dfrac{d^2T}{dt^2}$ also).

Thus we can write $\dfrac{1}{X}\dfrac{d^2X}{dx^2} = -\dfrac{1}{T}\dfrac{d^2T}{dt^2} = \lambda$

i.e. $\dfrac{1}{X}\dfrac{d^2X}{dx^2} = \lambda$ and $-\dfrac{1}{T}\dfrac{d^2T}{dt^2} = \lambda$

$\dfrac{d^2X}{dx^2} - \lambda X = 0$ and $\dfrac{d^2T}{dt^2} + \lambda T = 0$

The problem of solving the original differential equation (12.1) has now been reduced to that of solving two ordinary differential equations. Write down the general solutions of these when

(i) $\lambda = k^2$

(ii) $\lambda = -k^2$,

where k is real.

****

13A

(i) $X = A_1 e^{kx} + B_1 e^{-kx}$

   $T = A_2 \cos kt + B_2 \sin kt$

(ii) $X = A_1 \cos kx + B_1 \sin kx$

   $T = A_2 e^{kt} + B_2 e^{-kt}$

FOR TECHNOLOGISTS

FRAME 14

For it to be possible for y to equal sin t when x = 0, the first set of solutions in FRAME 13A must be chosen.

Thus $y = (A_1 e^{kx} + B_1 e^{-kx})(A_2 \cos kt + B_2 \sin kt)$

Does the condition $y \to 0$ as $x \to \infty$ tell us anything about any of the constants?

****

14A

Yes. $A_1 = 0$, as otherwise $y \to \infty$ as $x \to \infty$.

FRAME 15

Now we are left with

$$y = B_1 e^{-kx}(A_2 \cos kt + B_2 \sin kt)$$
$$= e^{-kx}(A \cos kt + B \sin kt) \quad \text{where} \quad A = B_1 A_2 \text{ and } B = B_1 B_2$$

Also, y = sin t when x = 0.

What further information does this give us about the constants?

****

15A

$A = 0, \quad k = 1 \quad \text{and} \quad B = 1$

$\therefore$ *The solution is* $y = e^{-x} \sin t$.

FRAME 16

In FRAME 2 we derived a partial differential equation giving the displacement of a tightly stretched vibrating string, i.e.

$$\frac{\partial^2 y}{\partial t^2} = c^2 \frac{\partial^2 y}{\partial x^2}$$

Suppose it is required to find y in terms of x and t if the motion is started

FRAME 16 continued

by displacing a stretched string of length ℓ into the form $y = a \sin \pi x/\ell$, and then releasing it.

Taking $y = XT$ as a trial solution, proceed to the stage corresponding to (12.2) in the previous example.

****

16A

$$X \frac{d^2T}{dt^2} = c^2 T \frac{d^2X}{dx^2}$$

$$\frac{1}{T} \frac{d^2T}{dt^2} = \frac{c^2}{X} \frac{d^2X}{dx^2} \quad (\text{or} \quad \frac{1}{c^2T} \frac{d^2T}{dt^2} = \frac{1}{X} \frac{d^2X}{dx^2})$$

FRAME 17

For the same reason as in the previous example, we can now write

$$\frac{1}{T} \frac{d^2T}{dt^2} = \frac{c^2}{X} \frac{d^2X}{dx^2} = \lambda \quad \text{(a constant)} \quad (17.1)$$

Which sign should be given to λ in this case and why?

****

17A

λ *must be negative as the solution of* $\frac{c^2}{X} \frac{d^2X}{dx^2} = \lambda$ *has to be trigonometric in order to accommodate the condition* $y = a \sin \pi x/\ell$ *when* $t = 0$.

FRAME 18

To ensure that λ is negative, put $\lambda = -k^2$ where k is real. (17.1) then yields

$$\frac{1}{T} \frac{d^2T}{dt^2} = -k^2 \quad \text{and} \quad \frac{c^2}{X} \frac{d^2X}{dx^2} = -k^2$$

Now find the general solutions of these two equations.

****

FOR TECHNOLOGISTS

18A

$$T = A_1 \cos kt + B_1 \sin kt$$
$$X = A_2 \cos \frac{kx}{c} + B_2 \sin \frac{kx}{c}$$

FRAME 19

Thus $y = \left(A_2 \cos \frac{kx}{c} + B_2 \sin \frac{kx}{c}\right)\left(A_1 \cos kt + B_1 \sin kt\right)$

The first condition is that when $x = 0$, $y = 0$ for all t.
What information does this give about any of the constants?

****

19A

$A_2 = 0$

FRAME 20

Now, replacing $B_2 A_1$ by A and $B_2 B_1$ by B, we can write
$$y = \sin \frac{kx}{c}\left(A \cos kt + B \sin kt\right).$$
The next condition is that when $t = 0$, $\frac{\partial y}{\partial t} = 0$ for all x as the string is initially at rest.

Use this to find further information about the constants.

****

20A

$$\frac{\partial y}{\partial t} = \sin \frac{kx}{c}(-Ak \sin kt + Bk \cos kt)$$
$B = 0$.

FRAME 21

Now $y = A \sin \frac{kx}{c} \cos kt$

The final condition is that when $t = 0$, $y = a \sin \pi x/\ell$.
Use this to write down the required solution for y.

****

$A = a$

$\dfrac{k}{c} = \dfrac{\pi}{\ell}$

$y = a \sin \dfrac{\pi x}{\ell} \cos \dfrac{\pi c t}{\ell}$

FRAME 22

As another example we will consider the following problem:

The current i in a cable satisfies the equation

$$\dfrac{\partial^2 i}{\partial x^2} = \dfrac{1}{c} \dfrac{\partial i}{\partial t} + i$$

A solution of this equation is required which will satisfy the conditions $i = 0$ when $x = \ell$, and $\dfrac{\partial i}{\partial t} = -ae^{-2ct}$ when $x = 0$, both for all values of t.

First, let $i = XT$ and show that it is possible to separate the variables to give the ordinary differential equations

$$\dfrac{d^2 X}{dx^2} - \lambda X = 0$$

and $\dfrac{dT}{dt} + k(1 - \lambda)T = 0$

****

22A

$T \dfrac{d^2 X}{dx^2} = \dfrac{X}{c} \dfrac{dT}{dt} + XT$

$\dfrac{1}{X} \dfrac{d^2 X}{dx^2} = \dfrac{1}{cT} \dfrac{dT}{dt} + 1 = \lambda$

giving $\dfrac{d^2 X}{dx^2} - \lambda X = 0$ \hfill (22A.1)

and $\dfrac{dT}{dt} + c(1 - \lambda)T = 0$ \hfill (22A.2)

FOR TECHNOLOGISTS

FRAME 23

Now complete the solution.

HINT: First solve equation (22A.2); the value of λ should then be apparent from the boundary conditions.

****

23A

The solution of (22A.2) is
$$T = Ae^{-c(1-\lambda)t}$$
The condition $\frac{\partial i}{\partial t} = -ae^{-2ct}$ *when* $x = 0$ *requires the exponent in T to be* $-2ct$.

$\therefore -c(1 - \lambda) = -2c$

$\lambda = -1$

Thus, $T = Ae^{-2ct}$

and (22A.1) becomes $\frac{d^2 X}{dx^2} + X = 0$

giving $X = B \cos x + C \sin x$

$\therefore i = e^{-2ct}(B_1 \cos x + C_1 \sin x)$

$\frac{\partial i}{\partial t} = -2ce^{-2ct}(B_1 \cos x + C_1 \sin x)$

Now, $\frac{\partial i}{\partial t} = -ae^{-2ct}$ *when* $x = 0$

$\therefore -a = -2cB_1$

$\therefore B_1 = \frac{a}{2c}$

$i = e^{-2ct}(\frac{a}{2c} \cos x + C_1 \sin x)$

Also $i = 0$ *when* $x = \ell$

$\therefore 0 = \frac{a}{2c} \cos \ell + C_1 \sin \ell$

$C_1 = -\frac{a}{2c} \cot \ell$

$i = \frac{a}{2c} e^{-2ct}(\cos x - \cot \ell \sin x)$

You will see that this can be written in the form
$$i = \frac{ae^{-2ct} \sin (\ell - x)}{2c \sin \ell}$$

FRAME 24

Examples involving the use of Fourier Series

We shall now consider the solution of equation (12.1)

$$\text{i.e.} \quad \frac{\partial^2 y}{\partial x^2} + \frac{\partial^2 y}{\partial t^2} = 0$$

which is valid for $0 \leq x \leq \pi$ for the following boundary conditions:

(a) $y = 0$ when $x = 0$ $\bigg\}$ for all t
(b) $y = 0$ when $x = \pi$

(c) $y \to 0$ when $t \to \infty$

(d) $y = 1$ when $t = 0$ for $0 < x < \pi$

Separation of the variables (see FRAMES 12 and 13) yields

$$\frac{d^2 X}{dx^2} - \lambda X = 0 \quad \text{and} \quad \frac{d^2 T}{dt^2} + \lambda T = 0$$

Which set of solutions in 13A is compatible with condition (c)?
Does this condition give you any information about the constants in the set you choose?

****

24A

Set (ii) with $A_2 = 0$ (assuming k positive).
Then $T = B_2 e^{-kt}$ which $\to 0$ as $t \to \infty$.

FRAME 25

Thus $y = e^{-kt}(A \cos kx + B \sin kx)$

Now apply condition (a).

****

25A

$A = 0$

FOR TECHNOLOGISTS

FRAME 26

We now have
$$y = Be^{-kt} \sin kx \qquad (26.1)$$

Applying condition (b) gives $0 = B \sin k\pi$.

What can you conclude about k?

****

26A

k must be an integer. (*B = 0 would give the solution* $y = 0$, *which would contradict condition (d).*)

FRAME 27

The only condition not used so far is (d). Substituting it in (26.1) gives
$$1 = B \sin kx \quad \text{for all x between 0 and } \pi.$$

This is impossible as B and k are constants, so a solution of the form (26.1) cannot be made to satisfy condition (d). To overcome this difficulty we proceed as follows.

It is known from 26A that k must be an integer, and so
$$y = B_1 e^{-t} \sin x, \quad y = B_2 e^{-2t} \sin 2x, \quad y = B_3 e^{-3t} \sin 3x, \quad \text{etc.}$$
are all solutions of $\frac{\partial^2 y}{\partial x^2} + \frac{\partial^2 y}{\partial t^2} = 0$ which satisfy conditions (a), (b) and (c).

A more general solution would be
$$y = B_1 e^{-t} \sin x + B_2 e^{-2t} \sin 2x + B_3 e^{-3t} \sin 3x + \ldots \qquad (27.1)$$

Verify that this solution satisfies conditions (a), (b) and (c).

FRAME 28

Applying condition (d) to this solution gives

$$1 = B_1 \sin x + B_2 \sin 2x + B_3 \sin 3x + \ldots \quad \text{for } 0 < x < \pi$$

We now have a feasible solution if the R.H.S. of this equation is the half-range Fourier sine series for 1,

$$\text{i.e. if } B_n = \frac{2}{\pi} \int_0^\pi 1 \cdot \sin nx \, dx$$

$$= \frac{2}{\pi} \left[-\frac{1}{n} \cos nx \right]_0^\pi$$

$$= \frac{2}{n\pi}(1 - \cos n\pi)$$

$$= \begin{cases} 0 & \text{if } n \text{ is even} \\ \dfrac{4}{n\pi} & \text{if } n \text{ is odd} \end{cases}$$

(A detailed treatment of Fourier Series is given in Programme 1 of this volume.)

The solution (27.1) then becomes

$$y = \frac{4}{\pi}(e^{-t} \sin x + \frac{1}{3} e^{-3t} \sin 3x + \frac{1}{5} e^{-5t} \sin 5x + \ldots)$$

FRAME 29

A stretched string of length ℓ, fixed at both ends, is set oscillating by displacing the mid-point a distance a perpendicular to the length of the string and releasing it from rest. The differential equation of the motion, derived in FRAME 2, is

$$\frac{\partial^2 y}{\partial t^2} = c^2 \frac{\partial^2 y}{\partial x^2}$$

First of all, write down the four boundary conditions which the solution has to satisfy.

****

FOR TECHNOLOGISTS

29A

(i) $y = 0$ when $x = 0$ for all t

(ii) $y = 0$ when $x = \ell$ for all t

(iii) $\frac{\partial y}{\partial t} = 0$ when $t = 0$ for all x

(iv) $y = 2ax/\ell$ from $x = 0$ to $x = \tfrac{1}{2}\ell$
      $y = (2a/\ell)(\ell - x)$ from $x = \tfrac{1}{2}\ell$ to $x = \ell$ $\Big\}$ when $t = 0$
      as the shape of the string when released is

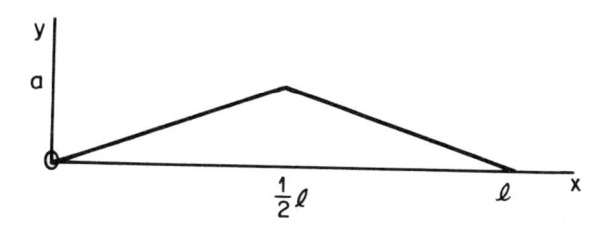

FRAME 30

Separation of the variables (see FRAMES 16 and 17) gives

$$\frac{d^2 T}{dt^2} - \lambda T = 0 \quad \text{and} \quad \frac{d^2 X}{dx^2} - \frac{\lambda}{c^2} X = 0$$

What must be the sign of λ?

****

30A

Negative.

For the string to oscillate, the solution must be trigonometric in t. Also, condition (iv) suggests that a Fourier series will be needed and this implies that the solution must be trigonometric in x.

FRAME 31

Putting $\lambda = -k^2$ and using conditions (i) and (iii), derive the form of solution $y = A \sin \frac{kx}{c} \cos kt$, and show that condition (ii) requires $k = n\pi c/\ell$, where n is an integer.

****

31A

The derivation of the solution $y = A \sin \frac{kx}{c} \cos kt$ *was given in FRAMES 18 – 21*

$y = 0$ when $x = \ell$ gives $0 = A \sin \frac{k\ell}{c} \cos kt$

∴ $\frac{k\ell}{c}$ *must be a multiple of* π

i.e. $\frac{k\ell}{c} = n\pi$ *where n is an integer.*

FRAME 32

The solution is now of the form
$$y = A \sin \frac{n\pi x}{\ell} \cos \frac{n\pi ct}{\ell}$$
or, more generally, $y = A_1 \sin \frac{\pi x}{\ell} \cos \frac{\pi ct}{\ell} + A_2 \sin \frac{2\pi x}{\ell} \cos \frac{2\pi ct}{\ell} + \ldots$
using the ideas given in FRAME 27.

Putting $t = 0$ gives $y = A_1 \sin \frac{\pi x}{\ell} + A_2 \sin \frac{2\pi x}{\ell} + \ldots$
and this must represent the function defined in condition (iv) for the range $0 < x < \ell$.

Find the coefficients of the terms in the half-range Fourier sine series for this function and complete the solution.

****

32A

$$A_n = \frac{2}{\ell} \left\{ \int_0^{\frac{1}{2}\ell} \frac{2ax}{\ell} \sin \frac{n\pi x}{\ell} dx + \int_{\frac{1}{2}\ell}^{\ell} \frac{2a}{\ell}(\ell - x) \sin \frac{n\pi x}{\ell} dx \right\}$$

$$= \frac{4a}{\ell^2} \left\{ \left[-\frac{\ell x}{n\pi} \cos \frac{n\pi x}{\ell} + \frac{\ell^2}{n^2\pi^2} \sin \frac{n\pi x}{\ell} \right]_0^{\frac{1}{2}\ell} \right.$$

$$\left. + \left[-\frac{\ell(\ell - x)}{n\pi} \cos \frac{n\pi x}{\ell} - \frac{\ell^2}{n^2\pi^2} \sin \frac{n\pi x}{\ell} \right]_{\frac{1}{2}\ell}^{\ell} \right\}$$

$$= \begin{cases} 0 & \text{if } n \text{ is even} \\ \frac{8a}{\pi^2 n^2} & \text{if } n = 1, 5, 9 \text{ etc.} \\ -\frac{8a}{\pi^2 n^2} & \text{if } n = 3, 7, 11 \text{ etc.} \end{cases}$$

$$y = \frac{8a}{\pi^2} \left[\sin \frac{\pi x}{\ell} \cos \frac{\pi ct}{\ell} - \frac{1}{3^2} \sin \frac{3\pi x}{\ell} \cos \frac{3\pi ct}{\ell} + \frac{1}{5^2} \sin \frac{5\pi x}{\ell} \cos \frac{5\pi ct}{\ell} \ldots \right]$$

FOR TECHNOLOGISTS 2:25

FRAME 33

Miscellaneous Examples

In this frame a collection of miscellaneous examples is given for you to try. Answers are supplied in FRAME 34, together with such working as is considered helpful.

1. Find a solution of $\dfrac{\partial^2 y}{\partial x \partial t} = \sin t$ which is such that $y = 2 + t$ when $x = 0$, and $y = 2$ when $t = 0$.

2. A metal rod AB, length ℓ, is placed with A at $(0,0)$ and B at $(\ell,0)$. The temperature θ satisfies the equation

$$\dfrac{\partial \theta}{\partial t} = k \dfrac{\partial^2 \theta}{\partial x^2} \qquad \text{where} \quad k > 0.$$

At time t the end A has temperature $\theta_o e^{-\omega t}$ ($\omega > 0$) while B is always kept at zero temperature. Assuming a solution of the type $\theta = X e^{-\omega t}$, where X is a function of x only, show that the temperature at the mid-point of the rod is

$$\tfrac{1}{2}\theta_o e^{-\omega t} \sec \tfrac{1}{2}\sqrt{\tfrac{\omega}{k}}\,\ell .$$

NOTE: Sometimes the form of the solution is partly known or given. In this example $\theta = X e^{-\omega t}$ is taken as the trial solution rather than the more general $\theta = XT$. Only one ordinary differential equation (that for X in this case) then arises.

3. If ϕ is the temperature at any point in a plane the equation for steady heat flow when expressed in polar coordinates is

$$\dfrac{\partial^2 \phi}{\partial r^2} + \dfrac{1}{r}\dfrac{\partial \phi}{\partial r} + \dfrac{1}{r^2}\dfrac{\partial^2 \phi}{\partial \theta^2} = 0$$

Find, by the method of separation of variables, a solution to this equation, the solution to involve θ trigonometrically. If $\phi = 0$ when $\theta = 0$ for all values of r and $\phi \to 0$ as $r \to \infty$ for all values of θ, show that the solution reduces to the form

$$\phi = C r^{-n} \sin n\theta$$

(HINT: In the ordinary differential equation for R in terms of r, either use the trial solution $R = a r^m$, where m is to be determined, or substitute $r = e^u$.)

FRAME 33 continued

4. A rod of length ℓ radiates heat according to the law

$$\frac{\partial \phi}{\partial t} = K \frac{\partial^2 \phi}{\partial x^2} - h\phi$$

h and K being positive constants. By means first of the substitution $\phi = e^{-ht} W$, where W is a function of x and t, and then separation of variables, find a solution of this equation given the conditions

    (a) $\phi = 0$ when $x = 0$,

    (b) $\phi = 0$ when $x = \ell$,

    (c) $\phi \to 0$ as $t \to \infty$.

5. The equation of motion for small transverse horizontal oscillations of a thin rod is

$$m \frac{\partial^2 y}{\partial t^2} + EI \frac{\partial^4 y}{\partial x^4} = 0.$$

Find a solution of this equation if it is known to be of the form $y = XT$ and T involves real trigonometric functions only. If the rod is of length ℓ and is clamped horizontally at both ends, show that $\cos n\ell \cosh n\ell = 1$ where $n^4 = mp^2/EI$ and $2\pi/p$ is the period of the oscillation.

6. Find a solution of $\frac{\partial^2 y}{\partial x^2} + c^2 \frac{\partial^2 y}{\partial t^2} = 0$ that satisfies the conditions:

    (a) $y = 0$ when $x = 0$

    (b) $y \to 0$ when $t \to \infty$

    (c) $y = a \sin 2cx + 2a \sin cx$ when $t = 0$.

7. For an insulated rod of constant cross-sectional area, the heat conduction equation is $\frac{\partial \theta}{\partial t} = h^2 \frac{\partial^2 \theta}{\partial x^2}$. A rod of length ℓ has one end (at $x = 0$) kept at $0°C$ and the other (at $x = \ell$) is kept at $100°C$ until steady state conditions prevail. The temperature of the hot end is suddenly reduced to zero. Find the expression for the temperature θ as a function of x and t, measuring t from the time when the end at $x = \ell$ has its temperature changed.

FOR TECHNOLOGISTS   2:27

FRAME 33 continued

8. In the steady flow of heat in the x,y plane, the temperature θ satisfies
    $\frac{\partial^2 \theta}{\partial x^2} + \frac{\partial^2 \theta}{\partial y^2} = 0$. A square plate has its edges along the lines $x = 0$,
    $x = \pi$, $y = 0$, $y = \pi$. The edges along $x = 0$ and $x = \pi$ are kept at
    zero temperature. The edge $y = 0$ is insulated and the edge $y = \pi$ is
    kept at $\theta = 100$. Show that
    $$\theta = \frac{400}{\pi}\left(\frac{\sin x \cosh y}{\cosh \pi} + \frac{\sin 3x \cosh 3y}{\cosh 3\pi} + \frac{\sin 5x \cosh 5y}{\cosh 5\pi} + \ldots\right)$$

9. The potential v at a distance x along a certain cable of length ℓ
    satisfies the equation $\frac{\partial^2 v}{\partial x^2} = rc \frac{\partial v}{\partial t}$ where r is the resistance/unit
    length and c the capacitance/unit length, both r and c being constant.
    The cable is at a uniform potential V throughout its length but at $t = 0$
    both ends are suddenly reduced to zero potential. Find

    (i) v as a function of x and t

    (ii) the total electrostatic energy at any time in the cable in
         terms of V, R, C and t where R and C are its total
         resistance and capacitance.
         (The E.S. energy at any time is $\int_0^\ell \tfrac{1}{2}cv^2\, dx$.)

10. A sheet of copper at $1080°C$ is suddenly immersed in a bath of oil at its
    boiling point, which is $180°C$. The sheet is 2 cm thick. Assuming that
    the sheet is of large enough dimensions for the heat losses from the four
    thin edges to be ignored, the temperature θ at a distance x from one
    surface of the sheet after time t satisfies the partial differential
    equation
    $$\frac{\partial \theta}{\partial t} = \alpha \frac{\partial^2 \theta}{\partial x^2}$$
    where α is the thermal diffusivity (for copper this is $0 \cdot 88$ cm^2/s).
    Calculate the temperature at the centre of the sheet after 2 seconds.
    (HINT: Substitute $\theta = \psi + 180$ in the differential equation and then
           find a solution for ψ to fit the boundary conditions.)

Answers to Miscellaneous Examples

1. $y = -x \cos t + F(t) + g(x)$

    $y = 2 + t$ when $x = 0$ ∴ $2 + t = F(t) + g(0)$

    $y = 2$ when $t = 0$ ∴ $2 = -x + F(0) + g(x)$

    i.e. $F(t) + g(0) = 2 + t$

    $F(0) + g(x) = 2 + x$

    One pair of functions is $F(t) = 2 + t$, $g(x) = x$.

    ∴ $y = -x \cos t + 2 + t + x$.

    More generally we could write $F(t) = c + t$ (c a constant) and $g(x) = x + 2 - c$. The same y would result.

2. Let $\theta = Xe^{-\omega t}$, then

    $$k \frac{d^2 X}{dx^2} + \omega X = 0$$

    ∴ $\theta = e^{-\omega t} \left(A \cos \sqrt{\frac{\omega}{k}} x + B \sin \sqrt{\frac{\omega}{k}} x \right)$

    $= e^{-\omega t} (A \cos px + B \sin px)$ if we write p for $\sqrt{\frac{\omega}{k}}$

    When $x = 0$, $\theta = \theta_o e^{-\omega t}$

    ∴ $\theta_o e^{-\omega t} = e^{-\omega t} A$ or $A = \theta_o$

    $\theta = e^{-\omega t} (\theta_o \cos px + B \sin px)$

    When $x = \ell$, $\theta = 0$

    ∴ $\theta_o \cos p\ell + B \sin p\ell = 0$ or $B = -\theta_o \cot p\ell$

    ∴ $\theta = \theta_o e^{-\omega t} (\cos px - \cot p\ell \sin px)$

    $= \theta_o e^{-\omega t} \sin p(\ell - x) \csc p\ell$

    When $x = \tfrac{1}{2}\ell$, $\theta = \theta_o e^{-\omega t} \sin \tfrac{1}{2} p\ell \csc p\ell$

    $= \tfrac{1}{2} \theta_o e^{-\omega t} \sec \tfrac{1}{2} p\ell$

    $= \tfrac{1}{2} \theta_o e^{-\omega t} \sec \tfrac{1}{2} \sqrt{\frac{\omega}{k}} \ell$

FRAME 34 continued

3. Let $\phi = RT$ (here, T is being used for a function of θ only), then

$$r^2 \frac{d^2R}{dr^2} + r \frac{dR}{dr} - \lambda R = 0 \quad \text{and} \quad \frac{d^2T}{d\theta^2} + \lambda T = 0$$

As θ is trigonometric, put $\lambda = k^2$, then

$$r^2 \frac{d^2R}{dr^2} + r \frac{dR}{dr} - k^2 R = 0 \quad \text{and} \quad \frac{d^2T}{d\theta^2} + k^2 T = 0$$

The latter gives $T = A \cos k\theta + B \sin k\theta$.

The former gives either

(i) $m^2 - k^2 = 0$, when the trial solution is used, leading to

$$m = \pm k$$

and $R = A_1 r^k + B_1 r^{-k}$

or

(ii) $\frac{d^2R}{du^2} - k^2 R = 0$ on putting $r = e^u$.

$$R = A_1 e^{ku} + B_1 e^{-ku}$$
$$= A_1 r^k + B_1 r^{-k}$$

Hence $\phi = (A_1 r^k + B_1 r^{-k})(A \cos k\theta + B \sin k\theta)$

$\phi = 0$ when $\theta = 0$ requires $A = 0$

$$\phi = (A_2 r^k + B_2 r^{-k}) \sin k\theta$$

$\phi \to 0$ as $r \to \infty$ requires $A_2 = 0$

$$\phi = B_2 r^{-k} \sin k\theta$$

which, if $k = n$ and $B_2 = C$ is of the form

$$\phi = C r^{-n} \sin n\theta.$$

FRAME 34 continued

4. If $\phi = e^{-ht}W$, the given equation becomes

$$\frac{\partial W}{\partial t} = K \frac{\partial^2 W}{\partial x^2}$$

Let $W = XT$, then

$$\frac{dT}{dt} - \lambda KT = 0, \quad \frac{d^2X}{dx^2} - \lambda X = 0$$

Put $\lambda = -n^2$. If you have tried $\lambda = n^2$, you will have found that it is impossible to satisfy both conditions (a) and (b).

$$\phi = e^{-(h+n^2K)t}(A \cos nx + B \sin nx)$$

(c) is already satisfied

(a) gives $A = 0$

$$\phi = Be^{-(h+n^2K)t} \sin nx$$

(b) gives $n\ell = k\pi$ where k is an integer

$$\phi = Be^{-(h+k^2\pi^2K/\ell^2)t} \sin \frac{k\pi x}{\ell}$$

Insufficient information is given in this example for B and k to be found.

An alternative method of solution is to do a direct separation of variables on the original equation without first getting the simpler W equation.

5. Let $y = XT$, then

$$\frac{d^2T}{dt^2} - \lambda T = 0, \quad \frac{d^4X}{dx^4} + k^4\lambda X = 0 \quad \text{where} \quad k^4 = \frac{m}{EI}$$

Put $\lambda = -c^4$ as T is to be trigonometric, then

$$\frac{d^2T}{dt^2} + c^4T = 0, \quad \frac{d^4X}{dx^4} - k^4c^4X = 0$$

The former gives $T = A_1 \cos c^2t + B_1 \sin c^2t$ and the latter has auxiliary equation $\mu^4 - k^4c^4 = 0$ if the trial solution $X = Ae^{\mu x}$ is used.

FRAME 34 continued

i.e. $\mu = kc,\ -kc,\ ikc$ and $-ikc$

giving $X = Ae^{kcx} + Be^{-kcx} + C\cos kcx + E\sin kcx$

or $X = A\cosh kcx + B\sinh kcx + C\cos kcx + E\sin kcx$

which is probably the more useful form here in view of what we are asked to prove.

$$y = (A_1\cos c^2 t + B_1\sin c^2 t)(A\cosh kcx + B\sinh kcx + C\cos kcx + E\sin kcx)$$

The conditions are (a) when $x = 0,\ y = 0$
(b) when $x = \ell,\ y = 0$
(c) when $x = 0,\ \dfrac{\partial y}{\partial x} = 0$
(d) when $x = \ell,\ \dfrac{\partial y}{\partial x} = 0$

(a) gives $A + C = 0$.

(b) gives $A\cosh kc\ell + B\sinh kc\ell + C\cos kc\ell + E\sin kc\ell = 0$

$$\frac{\partial y}{\partial x} = (A_1\cos c^2 t + B_1\sin c^2 t)kc(A\sinh kcx + B\cosh kcx - C\sin kcx + E\cos kcx)$$

and so (c) gives $B + E = 0$.

(d) gives $A\sinh kc\ell + B\cosh kc\ell - C\sin kc\ell + E\cos kc\ell = 0$.

Substituting for C and E in the equations given by (b) and (d),

$$A(\cosh kc\ell - \cos kc\ell) + B(\sinh kc\ell - \sin kc\ell) = 0$$
$$A(\sinh kc\ell + \sin kc\ell) + B(\cosh kc\ell - \cos kc\ell) = 0$$

The condition for these to have a non-zero solution for A and B is

$$\begin{vmatrix} \cosh kc\ell - \cos kc\ell & \sinh kc\ell - \sin kc\ell \\ \sinh kc\ell + \sin kc\ell & \cosh kc\ell - \cos kc\ell \end{vmatrix} = 0 *$$

or $(\cosh kc\ell - \cos kc\ell)^2 - (\sinh^2 kc\ell - \sin^2 kc\ell) = 0$

or $\cosh kc\ell \cos kc\ell = 1$

Let $kc = n$ then $\cosh n\ell \cos n\ell = 1$

and $\dfrac{EI}{m} = \dfrac{c^4}{n^4}$ or $n^4 = \dfrac{mc^4}{EI} = \dfrac{mp^2}{EI}$ if $c^2 = p$.

The time period of the oscillations is $2\pi/c^2$, i.e. $2\pi/p$.

* If you are unfamiliar with this result, solve each of the preceding equations for A/B and equate the two values.

FRAME 34 continued

6. Let $y = XT$, then

$$\frac{d^2X}{dx^2} - \lambda X = 0 \quad \text{and} \quad c^2 \frac{d^2T}{dt^2} + \lambda T = 0$$

Put $\lambda = -k^2$ (X must be trigonometric)

then $y = (A \cos kx + B \sin kx)(A_1 e^{kt/c} + B_1 e^{-kt/c})$

Condition (a) requires $A = 0$, then

$$y = (A_2 e^{kt/c} + B_2 e^{-kt/c}) \sin kx$$

Condition (b) requires $A_2 = 0$, thus

$$y = B_2 e^{-kt/c} \sin kx$$

To satisfy (c) it is necessary to use two values of k. If $k = c$, a solution is $y = B_2 e^{-t} \sin cx$ and if $k = 2c$, another solution is $y = B_3 e^{-2t} \sin 2cx$.

Hence $y = B_2 e^{-t} \sin cx + B_3 e^{-2t} \sin 2cx$ is also a solution.

Now $y = a \sin 2cx + 2a \sin cx$ when $t = 0$

$\therefore$ $a \sin 2cx + 2a \sin cx \equiv B_2 \sin cx + B_3 \sin 2cx$

giving $B_2 = 2a$, $B_3 = a$.

Hence $y = 2ae^{-t} \sin cx + ae^{-2t} \sin 2cx$.

You will notice that this example is intermediate between that in FRAME 16 where one value of k is required and that in FRAME 24 where it must take an infinite number of values.

FOR TECHNOLOGISTS

FRAME 34 continued

7. Let $\theta = XT$, then

$$\frac{dT}{dt} - \lambda h^2 T = 0, \quad \frac{d^2 X}{dx^2} - \lambda X = 0$$

Put $\lambda = -n^2$ ($+ n^2$ would make $\theta \to \infty$ as $t \to \infty$)

then $\theta = e^{-h^2 n^2 t}(A \cos nx + B \sin nx)$

When $x = 0$, $\theta = 0$ $\therefore A = 0$

$\theta = B e^{-h^2 n^2 t} \sin nx$

When $x = \ell$, $\theta = 0$ $\therefore n\ell = k\pi$ where k is an integer

$\theta = B e^{-h^2 k^2 \pi^2 t/\ell^2} \sin \frac{k\pi x}{\ell}$

When $t = 0$, $\theta = 100x/\ell$ which requires the more general solution

$$\theta = b_1 e^{-h^2 \pi^2 t/\ell^2} \sin \frac{\pi x}{\ell} + b_2 e^{-4h^2 \pi^2 t/\ell^2} \sin \frac{2\pi x}{\ell}$$

$$+ b_3 e^{-9h^2 \pi^2 t/\ell^2} \sin \frac{3\pi x}{\ell} + \ldots$$

Then when $t = 0$, $\theta = b_1 \sin \frac{\pi x}{\ell} + b_2 \sin \frac{2\pi x}{\ell} + b_3 \sin \frac{3\pi x}{\ell} + \ldots$

For the required distribution

$$b_k = \frac{2}{\ell} \int_0^\ell \frac{100x}{\ell} \sin \frac{k\pi x}{\ell} dx$$

$$= \begin{cases} \dfrac{200}{k\pi} & \text{when k is odd} \\ -\dfrac{200}{k\pi} & \text{when k is even} \end{cases}$$

$$\theta = \frac{200}{\pi} \left(e^{-h^2 \pi^2 t/\ell^2} \sin \frac{\pi x}{\ell} - \frac{1}{2} e^{-4h^2 \pi^2 t/\ell^2} \sin \frac{2\pi x}{\ell} \right.$$

$$\left. + \frac{1}{3} e^{-9h^2 \pi^2 t/\ell^2} \sin \frac{3\pi x}{\ell} - \frac{1}{4} e^{-16h^2 \pi^2 t/\ell^2} \sin \frac{4\pi x}{\ell} + \ldots \right)$$

FRAME 34 continued

8. Let $\theta = XY$, then

$$\frac{d^2X}{dx^2} - \lambda X = 0, \qquad \frac{d^2Y}{dy^2} + \lambda Y = 0$$

Put $\lambda = -k^2$ (When $y = \pi$, $\theta = 100$. Insertion of this condition will leave a function of x. This will probably have to be a Fourier Series, thus requiring trigonometric terms. $\therefore$ λ is expected to be negative.)

$\theta = (A \cos kx + B \sin kx)(A_1 \cosh ky + B_1 \sinh ky)$

$\theta = 0$ when $x = 0$ $\therefore$ $A = 0$

so $\theta = \sin kx (A_2 \cosh ky + B_2 \sinh ky)$

$\frac{\partial \theta}{\partial y} = k \sin kx (A_2 \sinh ky + B_2 \cosh ky)$

$\frac{\partial \theta}{\partial y} = 0$ when $y = 0$ $\therefore$ $B_2 = 0$ (This is the mathematical interpretation of the edge $y = 0$ being insulated.)

Now $\theta = A_2 \sin kx \cosh ky$.

$\theta = 0$ when $x = \pi$ $\therefore$ $\sin k\pi = 0$ i.e. k is an integer

$\theta = 100$ when $y = \pi$ cannot be satisfied by the present solution so take the more general form

$\theta = a_1 \sin x \cosh y + a_2 \sin 2x \cosh 2y + a_3 \sin 3x \cosh 3y + \ldots$

$\theta = 100$ when $y = \pi$ now gives

$100 = a_1 \sin x \cosh \pi + a_2 \sin 2x \cosh 2\pi + a_3 \sin 3x \cosh 3\pi + \ldots$

$= c_1 \sin x + c_2 \sin 2x + c_3 \sin 3x + \ldots$

where $c_k = a_k \cosh k\pi$

This can be satisfied by a half-range Fourier sine series for 100 ($0 < x < \pi$)

Then $c_k = \frac{2}{\pi} \int_0^\pi 100 \sin kx \, dx$

$= \begin{cases} 0 & \text{if k is even} \\ \dfrac{400}{\pi k} & \text{if k is odd} \end{cases}$

$a_k = \begin{cases} 0 & \text{if k is even} \\ \dfrac{400}{\pi k \cosh k\pi} & \text{if k is odd} \end{cases}$

$\theta = \dfrac{400}{\pi} \left(\dfrac{\sin x \cosh y}{\cosh \pi} + \dfrac{\sin 3x \cosh 3y}{3 \cosh 3\pi} + \dfrac{\sin 5x \cosh 5y}{5 \cosh 5\pi} + \ldots \right)$

FRAME 34 continued

9. Let $v = XT$, then if $rc = p^2$
$$\frac{d^2 X}{dx^2} - \lambda h^2 X = 0, \quad \frac{dT}{dt} - \lambda T = 0$$
Put $\lambda = -n^2$ ($+n^2$ would make $v \to \infty$ as $t \to \infty$)
$$v = e^{-n^2 t}(A \cos pnx + B \sin pnx)$$

When $x = 0$, $v = 0$ $\therefore A = 0$
$$v = Be^{-n^2 t} \sin pnx$$

When $x = \ell$, $v = 0$ $\therefore pn\ell = k\pi$ where k is an integer
$$n = \frac{k\pi}{p\ell}$$
$$v = Be^{-k^2 \pi^2 t/p^2 \ell^2} \sin \frac{k\pi x}{\ell}$$

or, more generally,
$$v = b_1 e^{-\pi^2 t/p^2 \ell^2} \sin \frac{\pi x}{\ell} + b_2 e^{-4\pi^2 t/p^2 \ell^2} \sin \frac{2\pi x}{\ell} + \ldots$$

When $t = 0$,
$$v = b_1 \sin \frac{\pi x}{\ell} + b_2 \sin \frac{2\pi x}{\ell} + \ldots$$
but $v = V$ when $t = 0$
$$b_k = \frac{2}{\ell} \int_0^\ell V \sin \frac{k\pi x}{\ell} dx$$
$$= \begin{cases} 0 & \text{when } k \text{ is even} \\ \frac{4V}{k\pi} & \text{when } k \text{ is odd} \end{cases}$$
$$v = \frac{4V}{\pi}\left(e^{-\pi^2 t/p^2 \ell^2} \sin \frac{\pi x}{\ell} + \frac{1}{3} e^{-9\pi^2 t/p^2 \ell^2} \sin \frac{3\pi x}{\ell} \right.$$
$$\left. + \frac{1}{5} e^{-25\pi^2 t/p^2 \ell^2} \sin \frac{5\pi x}{\ell} + \ldots \right)$$

ES energy = $\int_0^\ell \frac{1}{2} cv^2 \, dx$ and only the terms involving the square of a sine contribute to the integral.

The mth term in v is $\dfrac{4V}{(2m-1)\pi} e^{-(2m-1)^2 \pi^2 t/p^2 \ell^2} \sin \dfrac{(2m-1)\pi x}{\ell}$

FRAME 34 continued

The energy in the cable due to this term is

$$\int_0^\ell \frac{1}{2} c \frac{16V^2}{(2m-1)^2\pi^2} e^{-2(2m-1)^2\pi^2 t/p^2\ell^2} \sin^2 \frac{(2m-1)\pi x}{\ell} dx$$

$$= \frac{8cV^2}{(2m-1)^2\pi^2} e^{-2(2m-1)^2\pi^2 t/p^2\ell^2} \int_0^\ell \sin^2 \frac{(2m-1)\pi x}{\ell} dx$$

$$= \frac{8cV^2}{(2m-1)^2\pi^2} e^{-2(2m-1)^2\pi^2 t/p^2\ell^2} \frac{1}{2}\ell$$

$$= \frac{4CV^2}{(2m-1)^2\pi^2} e^{-2(2m-1)^2\pi^2 t/RC}$$

Total energy at this time $= \dfrac{4CV^2}{\pi^2} \sum_{m=1}^{\infty} \dfrac{1}{(2m-1)^2} e^{-2(2m-1)^2\pi^2 t/RC}$.

10. $\dfrac{\partial \psi}{\partial t} = \alpha \dfrac{\partial^2 \psi}{\partial x^2}$

Let $\psi = XT$, then

$$\frac{dT}{dt} - \lambda \alpha T = 0 \quad \text{and} \quad \frac{d^2 X}{dx^2} - \lambda X = 0$$

Put $\lambda = -k^2$ (ψ will not increase with time).

$\psi = e^{-k^2 \alpha t}(A \cos kx + B \sin kx)$

$x = 0$, $\psi = 0$ $\quad \therefore \quad A = 0$

$\psi = B e^{-k^2 \alpha t} \sin kx$

$x = 2$, $\psi = 0$ $\quad \therefore \quad \sin 2k = 0$

$\qquad\qquad\qquad\qquad k = n\pi/2$ where n is an integer

$\psi = B e^{-n^2\pi^2 \alpha t/4} \sin \dfrac{n\pi x}{2}$

or, more generally,

$\psi = B_1 e^{-\pi^2 \alpha t/4} \sin \dfrac{\pi x}{2} + B_2 e^{-\pi^2 \alpha t} \sin \pi x + B_3 e^{-9\pi^2 \alpha t/4} \sin \dfrac{3\pi x}{2} + \ldots$

For $t = 0$, $\psi = B_1 \sin \dfrac{\pi x}{2} + B_2 \sin \pi x + B_3 \sin \dfrac{3\pi x}{2} + \ldots$

At $t = 0$, $\psi = 900$

$\therefore B_n = \int_0^2 900 \sin \dfrac{n\pi x}{2} dx$

FRAME 34 continued

$$= \begin{cases} 0 & \text{if n is even} \\ \dfrac{3600}{n\pi} & \text{if n is odd} \end{cases}$$

$$\psi = \frac{3600}{\pi}\left(e^{-\pi^2 \alpha t/4} \sin \frac{\pi x}{2} + e^{-9\pi^2 \alpha t/4} \sin \frac{3\pi x}{2} + \ldots\right)$$

When $x = 1$, $t = 2$, $\alpha = 0 \cdot 88$

$$\psi = \frac{3600}{\pi} e^{-4 \cdot 34} \quad + \text{negligible terms)}$$

$$\simeq 15$$

The temperature at the centre of the sheet after 2 seconds = 195°C.
